SAFETY TRAINING
METHODS

Safety Training Methods

JACK B. RE VELLE

Management Consultant
Orange, California

A WILEY-INTERSCIENCE PUBLICATION

JOHN WILEY & SONS

New York • Chichester • Brisbane • Toronto

Copyright © 1980 by John Wiley & Sons, Inc.

All rights reserved. Published simultaneously in Canada.

Reproduction or translation of any part of this work beyond that permitted by Sections 107 or 108 of the 1976 United States Copyright Act without the permission of the copyright owner is unlawful. Requests for permission or further information should be addressed to the Permissions Department, John Wiley & Sons, Inc.

Library of Congress Cataloging in Publication Data:

ReVelle, Jack B.
   Safety training methods.

   "A Wiley-Interscience publication."
   Includes bibliographical references and index.
   1. Safety education, Industrial.   I. Title.
HD7260.6.R48   363.1'17'07   80-16601
ISBN 0-471-07761-5

Printed in the United States of America

10 9 8 7 6 5 4 3 2 1

For Bren and Karen,
whose safety I cherish

Preface

Safety Training Methods is intended for "hands-on" use by persons who are responsible and have authority for initiating and providing complete safety training programs in their organizations.

The book is both detailed enough for the neophyte employee and supervisor and broad enough for the experienced manager to provide a much-needed reference when designing, implementing, and monitoring a safety training program for the company.

As indicated in the contents, the book begins by describing how a safety training program fits into a company. The remaining chapters lead the reader through a carefully selected variety of safety training topics. Whether the reader needs specific information about the who, what, how, when, and why of safety training, OSHA training requirements, training in safety recordkeeping, fire safety, hazard inspection, accident investigation, and medical and first aid, or evaluating safety training effectiveness, it's all here in an easy-to-find, easy-to-use checklist format.

Additional features include a dictionary of commonly used safety and health terms, a model safety program, a variety of readily applied checklists, and several listings of safety- and health-oriented organizations, associations, publications, and periodicals, as well as other valuable sources of useful information.

When the book has been completely read, the reader should be able to design and implement a total safety training program in his or her own organization.

I look forward to hearing that readers have successfully applied the facts and techniques contained in this text. My pleasure as the author of this book will be incomplete until I learn of its acceptance by its readers.

Jack B. Re Velle

Orange, California
August 1980

Contents

1 How to Develop a Game Plan for Safety   1

    Management Involvement, 1
    Organizational Structuring of Safety, 10
    Team Up for Safety Training, 11
    The Safety Training Program, 15
    Bridging the Job-Home Gap, 17
    Promotion and Communication, 18
    Anticipated Results, 19

2 Facts About Safety Training   21

    How and When to Provide Safety and Health Training, 22
    Who Conducts Safety Training Programs, 26
    How to Time Safety Training, 26
    Location of Safety Training Programs, 26
    Presentation of Safety Training Programs, 27
    How to Motivate for Safety, 31
    Initiating Self-Appraisal, 33

3 Instructional Sequence for Job Training   34

    New Employee Orientation, 34
    Job Safety Analysis, 35

Employee Analysis, 37
Training for Safety, 37
How to Prepare the Learner, 38
How to Demonstrate the Job, 38
Training Sequence Content, 39
Training Program Characteristics, 39
Teaching Aids, 40

4 OSHA Safety and Health Training Requirements 42

5 The Supervisor's Role in Safety Training 66

Provide Safety Climate, 66
Supervisory Actions, 69
Special Procedures, 69
Other Goals and Objectives, 70
Management's Overview of Safety Training, 72

6 Training in Safety Recordkeeping 74

Classifying Accidents, 74
Records Maintenance, 75
Injury Experience Measurement, 77
Internal Data Analysis, 77
External Comparisons, 78

7 How-To's in Specific Safety Training Techniques 79

Suiting Up for Safety, 79
Responding to Unsafe Acts, 100
Recognizing Unsafe Conditions: OSHA's Priority Hazards, 104
Eliminating Unsafe Conditions, 113
Accident Report Preparation, 115
Accident Investigation, 118

Contents   xi

8   Fire Safety Training   122

Causes of Industrial Fires, 122
Fire Prevention Program Components, 123
Fires and Extinguishers, 125
Guarding Against Electrical Fires, 132
Even a Light Bulb Is a Hazard, 132
Provide Proper Protection, 132
How to Handle Hazardous Gases, 133
Understanding Flashpoints, 134

9   OSHA and OSHA Inspection Training   135

Background, 135
Recent Developments, 136
Training Responsibilities, 139
Hazard Abatement and Appeal Procedures, 140
Imminent Danger, 140

10   Obstacle Course to Safety Training   142

Identification of Housekeeping Violations, 142
Unsafe Conditions Involving Mechanical or Physical Facilities, 144
Other Obstacles, 145

11   Hazard Inspection Training   147

Continuing and Scheduled Safety Observation Tour, 147
High Accident Areas and Operations Emphasized, 154
Planning a Formal Inspection for Safety and Health Hazards, 156
OSHA Standards Used as Inspection Guides, 159
Simulated Plant Inspection, 160

12 Medical and First Aid Training — 174

Program Objectives, 174
Personnel Selection and Training, 175
Medical Records and Facilities, 177
OSHA Medical Training Standards, 180
Industrial Hygiene Training, 181
A Cooperative Training Effort, 182

13 Evaluating Safety Training Effectiveness — 185

Measures of Safety, 185
Statistical Techniques, 204

Appendix — 224

Definitions, 224
Model Safety Program, 225
Service Organizations and Associations, 232
U.S. Department of Labor, Office of the Solicitor: Regional Offices, 233
U.S. Department of Labor, Bureau of Labor Statistics: Regional Offices, 235
Publications and Periodicals, 235
Specific Sources of Safety Information and Data Analysis, 237

Index — 239

SAFETY TRAINING METHODS

1
How to Develop a Game Plan for Safety

MANAGEMENT INVOLVEMENT

Management Statement of Safety Policy

Once a director of safety has been selected, appointed, and officed, his or her first responsibility is to review the existing safety policy and, if none exists, prepare a current statement on safety for the signature of a well-known and respected member of the top management team.

The vital components of a safety policy are letterhead paper, a title, several brief paragraphs, and the signatures and titles of top management. Without any one of these, the overall statement lacks the official appearance and impact of a policy letter.

The content of the letter should be concise; that is, it should be directly to the point, relevant, and clear to the average employee. Where possible, a comparison should be made between the quality of the company's product/service and the importance of employee safety and health to the company.

To attain maximum impact with this document, the widest possible distribution should be made. Several methods could be used simultaneously: a full-sized copy on every bulletin board in the organization, a copy to every employee with the pay check, a copy at the beginning of the company safety manual, a copy in the firm's newsletter or its equivalent, and display in any other location where multiple employee exposure is likely. Examples of actual safety policy letters are presented below.

MANAGEMENT STATEMENT OF POLICY

It is very important that management commit itself to a safety philosophy published for all employees to see and review. *Typical* statements of policy are given in Tables 1-1A and 1-1B.

Table 1-1A Statement of Policy: Example 1[a]

Company Letterhead
Date

Safety and Health Policy

People are our most important asset—their safety is our greatest responsibility.

It is the policy of our company that every employee is entitled to a safe and healthful place to work.

When a person enters the employ of our company, he or she has a right to expect to be provided with a proper place in which to work, as well as proper machines and tools with which to do the job, and that the employee will be able to devote his or her energies to the work without undue danger.

Only under such circumstances can the association between employee and employer be mutually profitable and harmonious. It is our desire and intention to provide a safe workplace, safe equipment, proper materials, and to establish and insist on safe methods and practices at all times.

It is a basic responsibility for all to make the SAFETY of human beings a part of their daily, hourly concern. This responsibility must be accepted by each one who conducts the affairs of the company, regardless of the capacity in which he or she function.

Employees are expected to use the SAFETY equipment provided. Rules of conduct and rules of SAFETY shall be observed. SAFETY equipment must not be destroyed or abused.

The joint cooperation of employees and management in the observance of this policy will provide safe working conditions and accident-free performance to our mutual advantage.

We consider the SAFETY of our personnel to be of first importance, and we ask your full cooperation in making this policy effective.

Signature Block
Chief Operating Officer

Signature Block
Plant Manager

[a]Copyright American Society for Personnel Administration, 1974. All rights reserved American Society for Personnel Administration, 19 Church St., Berea, OH 44017.

Table 1-1B   Statement of Policy: Example 2

Company Letterhead
Date

Safety and Health Policy

It is our policy that the safety and health of our employees is equal in importance to the quality of our many fine products. We take great pride in far exceeding the minimum safety and health standards established by the OSHAct of 1970.

In a continuing effort to provide our employees with the safest, healthiest possible working conditions, we have subscribed to numerous organizations and periodicals that keep us current on OSHA standards as well as the latest protective equipment.

Additionally, we have developed a safety and health manual, a copy of which is provided to each of our employees. We believe that as our employees recognize our all-out efforts on behalf of their personal well-being, they will strive even more vigorously to make their place of employment as accident free and as healthy as is humanly possible.

As new employees join our growing work force, we will ensure that each of them participates in a safety and health orientation program as a part of the traditional introductory program.

We look forward to cooperatively working toward an environment that offers our employees the best in a safe, healthful workplace.

Signature Block
Chief Operating Officer

Signature Block
Plant Manager

Organizing Safety Committees

Once a director of safety has a realistic safety policy, the next step is the development of the two primary safety committees. The *employee* safety committee and the *executive* safety committee constitute the essence of a company safety program.

The *employee* safety committee should have at least 9, but not more than 21 persons, representing all departments and all shifts. This composition and sizing is necessary to be certain that adequate representation of various interest groups is maintained.

When organizing the *employee* safety committee, it is vital that no one be forced to attend committee meetings. Persons may be appointed, selected, or elected, or they may volunteer, but no one should ever be required to be a member of the *employee* safety committee. Only those persons with a genuine interest in company safety and health should be encouraged to participate.

The organization of the *executive* safety committee should follow that of the *employee* safety committee. However even if a supervisor expresses lack of interest in this group, it is absolutely necessary that he or she participate, to encourage the support of the subordinates in an active, creative, and productive safety program.

Both committees should meet at least once each month on the same day and at the same time, to reinforce the regularity of safety as a part of the company's operation. Agendas and minutes should be used to expedite and facilitate meeting time utilization.

The responsibility and authority of a safety committee are normally defined by upper level management; and the committee functions within the scope of responsibility and authority outlined by management. Before the initial meeting, experienced leaders should delineate the extent of the safety committee's ability to act. This will help avoid future confrontations regarding the committee's legitimate areas of concern.

Departmental Safety Meetings

Interested members should be appointed or elected safety leader for each department and/or shift on a quarterly basis. Regular rotation of the responsibility should encourage more persons to think about everyone's safety, not just the individual's.

Deparmental safety leaders should convene meetings two to four times each month in conjunction with the appropriate supervisor, so as not to conflict with established work schedules.

Departmental safety leaders' responsibilities include the following:

Management Statement of Policy

- Calling together the employees in the department and telling them about individual safety records or suggestions. See Table 1-2.
- Discuss company-wide safety programs: contests, awarding of plaques, safety dinners. If possible, invite a representative of the safety department to give a brief talk.
- Posting of safety records on the bulletin board.
- Holding departmental contests.

Departmental safety meetings are excellent for reinforcing safety consciousness regarding rules, regulations, and procedures. Safety and fire protection rules, regulations, and procedures are designed to help employees

Table 1-2 Checklist for Safety Meetings

The following items are presented for your convenience as you prepare for the regularly scheduled company safety meeting on _____ :
(date)

No.	Action Required	Individual Responsible	Date of Action Completed
1.	Obtain list of persons expected to attend.		
2.	Notify all individuals of scheduled meeting in writing (at least two, but preferably five working days in advance).		
3.	Remind all individuals of scheduled meeting either in person or by phone (at least two, but preferably four hours in advance).		
4.	Obtain a listing of items to be discussed at meeting and prepare sufficient copies of agenda for all attendees.		
5.	Distribute one copy of meeting agenda to all individuals prior to meeting (at least one, but preferably two days in advance).		
6.	Determine availability of meeting room(s) for meeting and schedule space for desired length of time (at least one week in advance).		

Table 1-2 (*Continued*)

No.	Action Required	Individual Responsible	Date of Action Completed
7.	Prepare meeting room for use by neatly placing chairs, tables, paper, pencils, speakers' materials, name plates, audiovisual equipment (e.g., projectors, tape recorders, screens, tapes, films), and so on, where required (at least one hour in advance).		
8.	Schedule audiovisual specialist to operate equipment listed in item 7.		
9.	Transcribe a record of meeting including names of attendees, date and time of meeting, subjects discussed, and list of assigned projects, with name of person designated responsible and due date for each project.		
10.	Prepare a written record of each meeting, including all subjects listed in item 9, and distribute to all persons on list from item 1 (no later than one week after the meeting).		

avoid injury and prevent fires and explosions. It is imperative that these rules be promulgated to all employees either by posting on bulletin boards, a safety manual, and discussion at safety meetings or, better yet—through all these channels. Generally they can be divided into four groups.

- **General Plant Rules.**   These are safety and fire protection regulations that apply to all people in the plant. Usually they are formulated by the safety committee and issued over management's signature. Examples include smoking restrictions, prohibition of horseplay, traffic controls, and lockout procedures.
- **Area Rules.**   These apply to specific plant areas and are to be observed by all persons in these areas. Such rules usually prescribe the basic or minimum safety equipment to be worn in the area, such as eye protection, hard hats, and safety shoes.

- **Building Rules.** These may parallel the area rules in many respects; they are frequently more restrictive, however, and reflect specific hazards of a particular building.
- **Specific Rules.** Specific safety rules are usually an integral part of an operating or job procedure and pertain directly to the safe method of performing a specific task. Not only do they prescribe safe methods, but they specify also the required special protective clothing and equipment.

Periodic Review of Rules

Frequently changes in policies, processes, or procedures are introduced that require different rules and procedures. An annual review of all rules is a good way to keep them up to date and in mind.

It is suggested that group members receive safety statistics for their department for some convenient period—six months or a year. The leader may secure these data from the safety department.

Get group members to target an area for improvement during the next six months. Such areas, which should be listed on the chalkboard, may include the following:

- Breaking of safety rules.
- Weight-lifting hazards.
- Principal hazards in each department.
- Hand and finger injuries.
- Back injuries.
- Sanitation in the work area.
- Housekeeping.
- Storage of materials and equipment.
- Off-the-job accidents.
- Personal protective equipment.

Attempt to define goals in quantitative terms. It is not enough to say "improve housekeeping." Draw from the group some specific objectives, for example:

- Reduce area in which useless scrap is stored by 10%.
- Motivate each employee to turn in one suggestion on how to prevent injuries.

Get commitment from the group. "Let us each state how we will go about achieving our goal." The group may volunteer to take one or more of the following actions:

- Hold weekly safety meeting.
- Review safety rules.
- Make periodic safety inspection.

The group leader should point out that sometimes employees, in their desire to help set a good departmental record, simply do not report minor accidents or do not go for first aid. "We must watch out that this does not occur. The well-being of our employees is more important than making a good showing on paper."

Employee Safety Suggestion System

There should be some formal suggestion system for receiving employee safety ideas.

Appropriate forms should be available on which the employees can submit their ideas in writing. Having the suggestions forces the supervisor to take action one way or the other and keeps good ideas from falling by the wayside. It is bad enough if an employee makes a bona fide safety suggestion and never gets an answer from management. It is even worse when an employee makes a suggestion, no action occurs, and then an accident takes place involving the hazard in question. These problems can be avoided through the use of a viable safety suggestion system that is used by top management.

Stimulate the submission of safety suggestions. If some suggestions are not acceptable, fully explain why. Make certain that employees don't think their ideas are not adopted because of the cost to the company.

Give adequate advance notice of new safety measures, methods, and rules with an opportunity for employees to have some say in safety matters that will directly affect them.

Be on the lookout for ways in which individuals can participate directly in the accident prevention program, for example: as safety instructors or coaches, as safety committee members, as members of housekeeping teams, or as hazard inspectors.

The rules for brainstorming are given below. Explain them in a departmental safety meeting.

1. At first, judicial appraisal of an idea is ruled out. Criticism of ideas is withheld until later. No evaluation or comments, such as "this won't

Management Statement of Policy

work" or "we tried this before," are permitted. Brainstorming is "green light" thinking.
2. People in the group should let their minds wander. They should not be afraid of wild ideas. It is easier to "tame down" than to "think up."
3. The group should strive for quantity. The greater the number of ideas, the greater the likelihood of winners.
4. Combination and improvement is sought. In addition to contributing ideas of their own, participants should "hitchhike" on the ideas of others, suggesting how they can be turned into even better ideas or how two or more ideas can be joined to form still another idea.

To start off the brainstorming session, the leader may wish to mention a few of the following ideas used by companies:

- A bulletin board with a large reproduction of Sherlock Holmes was featured in a contest. Sherlock, in the form of an employee of the safety department and wearing a "detective" hat, visited each department for three days in a row and posted safety information. A gift certificate was awarded the employee who, when called on the phone at home, could answer a safety question based on information in the bulletins.
- Bright yellow hard hats were worn by an electric utility crew as they attended an Orange Bowl game. They wanted everyone to see that they were celebrating an all-time safety record.
- In a mine, crews competed to keep a black lantern trophy away from their districts. The lantern was labeled, "the light that doesn't shine," to indicate that work was not performed in a good manner. Points were given for safety infractions. The crew with the most points had to display the lantern.
- One company includes a safety pledge card with its safety instruction booklet. The card is signed by new employees and turned in to the supervisor, putting the employee on record as having pledged to make safety a habit.
- Another company distributes a souvenir keycase with a slogan KEYS—keep each year safe—as a safety reminder.
- A hat decal is given to each person who goes for a year without an accident ("1" for the first year, "2" for the second, etc.) At 10 years the employee gets a gold hat with a 10 year decal. Employees continue getting numbered decals as long as their records are unbroken.
- Safety calling cards are issued by foremen and supervisors to an employee observed performing an unsafe act. This person keeps the

card until he or she catches someone else working unsafely, at which time the card is passed on.
- A small safety box is placed conspicuously by the punch clock in a mill. Lettered on the outside:

>Inside this Box
>Is the Best Darn
>Accident Preventer
>In the Mill.

On opening the box, the curiosity seeker is confronted by a mirror—and his own face. The message reads: YOU!

ORGANIZATIONAL STRUCTURING OF SAFETY

Centralized Versus Decentralized Operations

Centralized

Active management and control of a company safety program may be vested in the chief executive, the general manager, or an experienced and qualified foreman who has both authority and status.

There are several advantages to safety inherent in small-scale operations, such as closer contact with the working force, more general acquaintance with the problems of the whole plant, and, frequently, less labor turnover.

The safety manager does have special problems with engineering and medical services. He or she is not likely to be in a position to hire full-time safety professionals or a full-time physician and/or nurse.

Decentralized

Organizations with scattered operations requiring relatively few employees, such as scattered construction sites, face special problems of organization. Their operations may be seasonal or intermittent, and there may not be a sufficiently stable working force to operate committees effectively. The local manager may need to adapt the safety program to local conditions, which may be quite variable.

Staff Versus Line Positions

The safety program is usually assigned to persons holding line positions in a small plant, and staff positions in a large plant. If a line official in a small plant has a safety function for portions of the plant over which he or she has no line authority, however, the safety asignment is considered to be a staff function. In a large plant the safety director and organization should have staff status and authority.

The exact determination of the organizational status of the safety staff must be decided by each firm in terms of its own operational problems, policies, and hazards.

Authority Versus Responsibility

Sometimes the safety professional is given authority that is usually limited to line officials. This authority is necessary to meet the responsibilities associated with the safety professional's position. Without the authority to act, the safety professional might not be able to fulfill his or her responsibilities.

On fast-moving and rapidly changing operations or those on which delayed action would endanger the lives of workers or others, it is not uncommon for the safety director to have authority to order immediate changes. Examples include construction and demolition work, fumigation, some phases of explosives manufacturing, and emergency work. Such authority, whenever and wherever granted, must be used with discretion, since the safety professional will be accountable for errors in judgment.

Matrix Versus Traditional Structure

Matrix organization (also called project management and program management) is most often used when a new product must be developed—for example, an organization must begin pilot production of a new type of engine, while maintaining production of existing models.

The primary advantage of a matrix organization, which is ordinarily a temporary supplement to a traditional structure, is that an objective can be achieved without expending the money and time required to develop a totally new organization. In many instances personnel can be drawn from within the parent organization without seriously impairing its efficiency (Figure 1-1).

The safety director with the dual responsibilities of both the traditional parent organization and the matrix appendage must understand the nature of matrix management to effectively perform the entire scope of his or her safety responsibilities.

TEAM UP FOR SAFETY TRAINING

Influencing People

The human relations movement, a post–World War II phenomenon, has as its primary focus the handling (treatment) of employees as human beings.

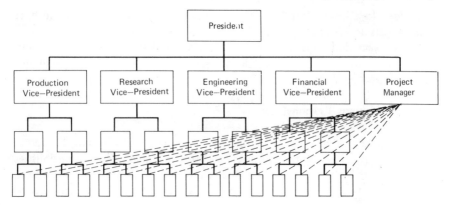

Figure 1-1 Matrix organization chart. From J. Clifton Williams, *Human Behavior in Organizations*, South-Western Publishing Company, Cincinnati, 1978, p. 44.

Most recent studies indicate that organizations that are most considerate of the feelings and concerns of their employees have the lowest accident frequency and accident severity rates. Therefore it is in the mutual best interests of companies and their employees to have supervisors who are skilled in the creation of organizational climates that are conducive to employee self-motivation. This level of supervisory skill is achieved through a judicious blend of training and experience. People are strongly influenced by the attitudes and actions of their supervisors.

Empirical studies indicate that some employees react negatively to even the most positive actions by their supervisors. Yet one of the responsibilities of leadership is the enforcement of discipline. Indeed, no matter how well-qualified supervisors may otherwise be, their fitness to lead is severely limited by their ability and willingness to impose necessary discipline on a timely basis. Remember, discipline is not just punishment, it is the creation of an organizational structure wherein all members knows exactly where and how they fit and precisely what is expected of them by both supervisors and peers. Most supervisors require extensive training to provide a high calibre of organizational discipline.

Supervision

The responsibilities of the first-line supervisor are many. Direction of the work force includes the following supervisory functions:

1. Setting goals.
2. Improving present work methods.

3. Delegating work.
4. Allocating manpower.
5. Meeting deadlines.
6. Controlling expenditures.
7. Following progress of work.
8. Evaluating employee performance.
9. Forecasting manpower requirements.
10. Supervising on-the-job training.
11. Reviewing employee performance.
12. Handling employee complaints.
13. Enforcing rules.
14. Conducting meetings.
    and last, but not least—
15. Increasing safety awareness.*

Supervisory understanding of the interrelationships of these responsibilities is a learned attribute. Organizations that expect their supervisors to offer a high quality of leadership to their employees must provide appropriate training and experiential opportunities to current supervisors and supervisory trainees alike.

Personality Factors in Safety

There is no doubt that from time to time, supervisors must become involved in the personal problems of their employees when those problems interfere with the work. Table 1-3 lists some of these problems.

The causes employee dissatisfaction must be well within the knowledge of supervisors; and they must also be able to take corrective measures. Qualified persons should be used to help supervisors to recognize and deal with employee personality factors that have an adverse impact on job achievement.

The influence of tension and stress as causes of failure to perform work safely is frequently noted by behaviorists. Supervisors must be constantly alert for symptoms of stress in their employees, to be able to take necessary and appropriate action to inhibit the likelihood of an accident. Table 1-4 lists some common symptoms of stress.

---

*"The Supervisor in 1975" by Blake D. Lewis, Jr. Reprinted with permission from *Personnel Journal,* Costa Mesa, CA, copyright September 1973.

Table 1-3 Employee Personal Problems[a]

Absenteeism	Irrationlity
Aggressiveness	Irresponsibility
Bizarre behavior	Irritability
Chronic rule breaking	Low productivity
Constant complaints	Memory lapses
Criticism of others	Perfectionism
Daydreaming	Poor judgment
Defensiveness	Preoccupation with health problems
Errors and mistakes	Procrastination
Excuse making	Rapid personality changes
Fearfulness	Self-criticism
Forgetfulness	Self-preoccupation
Guilt feelings	Shyness
Heavy drinking	Suspiciousness
Inability to concentrate	Tardiness
Inappropriate emotions	Unpredictable behavior
Indecisiveness	Use of narcotics
Indifference or apathy	Withdrawal
Inferiority feelings	

[a]J. Clifton Williams, *Human Behavior in Organizations*, South-Western Publishing Company, Cincinnati, 1978, p. 411. Reprinted with permission.

Program Promotion

Responsibility for communication of positive safety attitudes is spread from top management to the first-line supervisor. There is no greater negative safety influence than for employees to view members of company management failing to observe safety precautions that are enforced for the employees. The most graphic example is the nonuse by a manager of personal protective equipment (e.g., a hard hat or hearing protection), when employees are directed to make constant use of the equipment in the same area.

Communication occurs on several levels, sometimes without the sender or receiver realizing it has occurred. The sighting by an employee of a manager without a hard hat in a hard hat area weighs considerably more heavily on the future actions of that employee than do many safety notices on a bulletin

Table 1-4 Common Symptoms of Stress[a]

Anxiety or apprehensiveness	Hyperventilation
Chronically bad breath	Inability to concentrate
Chronic worrying	Inability to relax
Compulsive eating	Inability to sleep
Constant fatigue	Increased aggression
Constant inner tension	Increased defensiveness
Dependence on tranquilizers	Irritability and restlessness
Dizziness	Low energy level
Excessive smoking	Moodiness
Excessive use of alcohol	Pounding heart
Gastrointestinal disorders	Recurring headaches
Growing feelings of inadequacy	Sweaty hands or feet
High blood pressure	Temper outbursts

[a] J. Clifton Williams, *Human Behavior in Organizations,* South-Western Publishing Company, Cincinnati, 1978, p. 352. Reprinted with permission.

board. Managers and supervisors must learn that their actions do speak louder than words. They must learn to back their written and spoken communications with visual reinforcements.

Motivation of safe behavior is an abbreviated way of describing several steps that ultimately lead to employees being internally motivated, not coerced, toward safe behavior. First, persons with managerial and/or supervisory responsibilities must learn how behavior patterns are acquired. When this is understood, they should review the various behavior situations they encounter on a regular basis. Next they must choose appropriate behaviors to solve employee performance problems. When this is accomplished, it is necessary to encourage the desired employee practices. Finally, managers and supervisors must adopt known behavioral principles to their personal leadership styles.

THE SAFETY TRAINING PROGRAM

Foundation

The safety training program is pervasive in its scope in that its influence must be felt across and within the entire operation of a company.

Such a program has its foundation in personnel selection and orientation as well as in job-related training. Persons responsible for hiring and initial training of new employees must be alert to clues of negative safety attitudes and actions. It is important to take early steps either to correct these problems or to decide against employing individuals who appear to have poor safety attitudes.

Sometimes negative clues are well hidden in a persons' past employment record and are not brought out in reference checks. The last chance to identify a poor attitude is during on-the-job training. Trainers should be cognizant of attitudes and work habits that indicate the likelihood of problems in the area of safety.

Specialized Training

Once an individual is employed, he or she may be assigned certain responsibilities that are beyond the basic job requirements. Some areas of specialized training are as follows:

Accident investigation.
Accident report preparation.
Fire safety.
Hazard inspection.
Medical and first aid.
Personal protective equipment.
Powered equipment and vehicles.
Safety recordkeeping.
Specific disasters.

Appropriate trainers, facilities, and equipment must be provided on a timely basis to minimize interference with daily operational requirements.

Continuing Responsibilities

Provision of initial training in any or all areas of concern does not discharge a company's responsibility for employee training. It is necessary to maintain complete records of training (dates, subjects, names of those in attendance, exam grades, copies of exam, instructors' names, individual evaluations of trainees by trainers, etc.), so that scheduling of future training (refresher, advanced, specialized, etc.) can be accomplished logically.

BRIDGING THE JOB-HOME GAP

Impact on Employees

Off-the-job injuries have the same consequences to employees as those that occur on the job. These may include some or all of the following:

- Immediate loss of earning power.
- Reduced earning power resulting from permanent disability.
- Limited activity.
- Mental anguish, worry, and anxiety.
- Upset in normal living routine.
- Need to fill out accident reports and assist in investigation.
- Possibility of court appearance.
- Pain and discomfort.

Impact on Employers

Off-the-job injuries also influence employers' operations, as suggested below:

- Work delay while replacing an injured employee.
- Production loss due to absence of injured employee.
- Supervisor's time necessary to train replacement.
- Tool and material damage caused by inexperienced employee.
- Accident and medical benefit costs.
- Personnel department costs to replace injured employee.
- Wages paid to replacement employee.
- Cost of accident investigation and increase in insurance premiums.

Action Required

Discussions frequently reveal that employees are unaware of the impact of off-the-job injuries on their employers. Because of the obvious and extensive costs associated with off-the-job injuries, many employers are now training their employees on the safe performance of their activities while off the job. Some areas that can be included are as follows:

- Vehicle operation.
- Equipment selection.

- Equipment operation.
- Equipment maintenance.
- Equipment storage.
- Use of ladders.
- Building and facilities repair.

The extensiveness of such a program depends a great deal on the type of employees a company hires. No rules of thumb have been developed yet in this emerging area of concern.

PROMOTION AND COMMUNICATION

Telling and selling safety is a full-time responsibility. Promotional programs are useful in keeping employees interested. Activities may involve many media. Novelty or innovation often provides interest-getting appeal, which is important if the program is to do its job. However too bizarre an approach may mask the safety message and cause the program to lose much of its effectiveness. Any promotional campaign should have specific objectives, and the material used should be consistent with these objectives.

Some media for promoting safety include the following:

1. Awards.
2. Contests.
3. Films and tapes.
4. Handbooks.
5. Literature racks.
6. Meetings.
7. Meeting handouts.
8. Newsletters.
9. Payroll stuffers.
10. Telephone ("dialing for safety").
11. Posters.

Selected carefully and used with discrimination, posters can help your employees avoid hazards and unsafe acts and can help you have a more effective safety program. Here are some suggestions to help you get the most out of the safety posters you buy and use.

Change Them Often. A week is about the maximum time a poster should be displayed. Some organizations change poters evey two or three days.

Locate Them with Care. Put your posters up—either in poster frames or on bulletin boards—at strategic places, that is, where people pass, where they tarry, where they stop and read (near vending machines; at smoking stations; in lounges or cafeterias, tool rooms, conference or meeting rooms; near telephone booths; near time clocks; outside supervisors' offices).

Display One at a Time. Display one poster at one location at one time. There are exceptions, depending on the location and the particular display, but as a rule, don't mass posters—the attention factor is higher, and readability is better if the number of posters in one display is kept to a minimum.

Light Poster Display Well. Flashing lights and colored lights are effective. Frame posters. A frame does a lot for a poster—as it does for a picture. Don't use posters that are dog-eared or dirty.

Use Posters That Pertain to the Location at Which They Are Displayed. A poster featuring a model wearing high heels, for example, would be out of place at a location where no women are employed. Use specific posters at specific places; use general posters at general places.

Create Some Permanent Displays. Some subjects and some posters are so basic that permanent displays should be constructed around them. For example, certain fire protection, first aid, and rescue posters are important enough to warrant permanent display in some locations. In these cases the posters should be mounted on cardboard, shellacked, or placed in a glassfront poster box.

Use Special Notices. Also use poster locations for news bulletins, press clippings of accidents, and homemade posters and signs. Aside from the intrinsic value of such communications, they also draw people to poster locations and condition them to check those spots for interesting, timely, or amusing reading matter.

ANTICIPATED RESULTS

As work and time progress and employees around the company begin to "get the word" that the safety program is more than just a passing fancy of the president or the plant manager, it is reasonable to expect positive results. Whatever other benefits may be realized, the following interrelated objectives should represent the fulfillment of the company safety program.

1. One can logically expect to see reduced accident frequency rates, that is, the number of accidents that have occurred based on a standardized work year for 100 employees: 40 hours per week times 50 weeks per year per employee times 100 employees equals 200,000 worker-hours, the statistical basis used by the Occupational Safety and Health Administration (OSHA).
2. One can also expect to see lowered accident severity rates; this is the number of lost work days per recorded accident based on the 100-employee standardized work year.
3. As a direct result of items 1 and 2, the organization can anticipate cheaper workmen's compensation insurance premiums. The insurance carriers of this expensive coverage have demonstrated tremendous interest in assisting their clients to reduce their accident frequencies and accident severity rates so that their insurance payouts and the company's premiums can be significantly cut.
4. Other evidences of a substantially enhanced safety posture within an organization that are more difficult to quantify but are nevertheless the direct result of broadly based employee support of a safety program are improved housekeeping and expanded hazard reporting. The former can be quantified through the use of random sampling, and the latter can be ascertained by comparing totals of types of hazard report by the week, month, or quarter. Specific measures of evaluation of safety program effectiveness are addressed extensively in Chapter 13.

2

Facts About Safety Training

Before an employee can be expected to work safely, he or she must be shown the safe way to do the job.

A wide variety of training areas should be considered in most companies. Included in this list are training for mechanized equipment operators, first aid administrators (both basic and advanced), and accident investigators, training in fire hose location and operation, emergency evacuation drills (for both fire and bad weather), oxygen bottle operation (for first aid administrators only), and fire extinguisher operation, and training of new employees (part of orientation program) and maintenance personnel. See Figures 2-1 and 2-2.

These training programs should be scheduled in conjunction with supervisors' requirements to minimize operational interruptions. Upon completion of a training program, an examination (announced at the beginning of the program) should be administered and the grade recorded. Documentation of participation should be maintained by name, dates, total hours, grade on final examination, and so on to satisfy reviews by OSHA compliance officers.

To assure that high quality training is provided to all company personnel, the safety manager and other members of the management team should review the proposed course outline and materials. The course instructors should be knowledgeable personnel, from the company whenever possible, who are capable of logically communicating with a group of employees in their area of specialization.

In addition, it is possible to evaluate the quality of company training programs and personnel by observing correlations between examination scores and accident frequency and severity rates before and after training by department and by shift. This is discussed further in Chapter 13.

HOW AND WHEN TO PROVIDE SAFETY AND HEALTH TRAINING

It is important to know how and when to provide safety and health training. Such material is especially directed to employees performing new jobs to be sure that they are made aware of hazards and maintain this awareness.

The purpose of occupational safety and health training is to make employees aware of the safety and health hazards under which they operate during the working day, as well as to show them how to perform their jobs without endangering themselves or their fellow employees. The following quotations from the Occupational Safety and Health Act of 1970 (OSHAct) standards are *examples* of the emphasis placed on safety and health training by the act:

CFR 1910.178(1) Powered Industrial Trucks. Only trained and authorized operators shall be permitted to operate a powered industrial truck. Methods shall be devised to train operators in the safe operation of powered industrial trucks.

CFR 1910.217 (f)(1) Mechanical Power Presses. The employer shall train and instruct the operator in the safe method of work before starting work on any operation covered by this section.*

Before any employee starts a job that is new to him or her, it is the company's responsibility to provide safety and health training for that specific job. Thus safety and health training is required for a new employee, an employee who is changing jobs, or an employee who is about to start a newly created job or a job has been changed. The immediate supervisor is the best person to provide safety and health training for employees starting jobs new to them because this individual knows the most about the hazards involved in the jobs under his or her supervision.

Training an employee before work is started on a job affords a good opportunity to influence the formation of good work habits. With newly hired employees who are beginning their first jobs, appropriate training can help in the formation of safe and healthful work habits that will continue throughout

---

*Code of Federal Regulations, *Federal Register,* Washington, D.C.

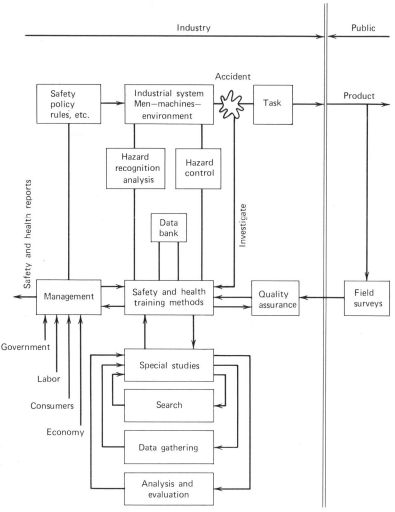

Figure 2-1 Basic functions of an effective safety program. Reprinted with permission from *Industrial Engineering Magazine.* Copyright American Institute of Industrial Engineers, Inc., 25 Technology Park/Atlanta, Norcross, GA 30092.

their entire working lives. With newly hired employees who have past work experience, an employer can get them off on the right foot. With experienced employees who have been assigned to new jobs or to newly created jobs, or to changed jobs within the organization, appropriate training can detect and correct unsafe or unhealthful work habits of which the employees may not even be aware. Providing safety and health training for employees starting new jobs new to them is an important supervisory function.

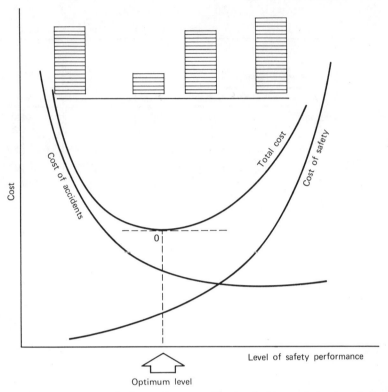

Figure 2-2 Optimum level of safety performance. Reprinted with permission from *Industrial Engineering Magazine.* Copyright American Institute of Industrial Engineers, Inc., 25 Technology Park/Atlanta, Norcross, GA 30092.

Safety and health training should include the following information:

- The rights and responsibilities of employees under the OSHAct, including conduct on the part of employee that may be necessary to comply with an OSHA standard.
- First aid and emergency procedures.
- The need to report all events resulting in injury, illness, or property damage, as well as near misses.
- The need to report hazardous conditions and practices.
- Company and departmental safety and health rules and their application to the employee's job.
- Specific hazards of the employee's job and how to avoid them.
- Procedures, precautions, safeguards, and personal protective equipment necessary to protect the employee from the hazards of the job.

Safety and health training for a new employee includes the four R'S: responsibilities, rights, rules, and regulations:

- Teaching a new employee his or her RESPONSIBILITIES under the OSHAct.
- Teaching a new employee his or her RIGHTS under the OSHAct.
- Teaching a new employee the general and departmental safety and health RULES of the company.
- Teach new employees starting jobs new to them about the applicable OSHAct safety and health REGULATIONS (standards).

Safety and health training should also include information about action to be taken in the following emergency situations:

- What to do if the employee or a fellow employee is seriously injured or develops severe symptoms of illness.
- How and when to evacuate the building rapidly.
- How and when to give a fire alarm to fellow employees, report a fire to the local fire department, and use portable fire extinguishers and other available fire fighting equipment.
- How and when to shut off electricity, gas, and other special hazards.

It is especially important for employees to know what to do in emergency situations because quick action is required.

Since the employer is responsible for correcting hazards and recording illnesses and injuries caused by occupational hazards, a good training program will teach employees to report such conditions and events. New employees should be told about the reporting requirements of the OSHAct. Since regulations promulgated under the OSHAct require employers to maintain specific information on all recordable injuries and illnesses, the need for employees to report all job-related injuries and illnesses that occur in their workplaces should be stressed.

It is a good idea to have employees report *all* injuries and illnesses so that management can decide whether an event is *recordable*. Then too, some "near misses" that are not actually recordable have come very close to being serious enough to count. Something can often be learned from a near miss that will help prevent the real event from happening.

The importance of prompt reporting of hazardous conditions or practices should be explained during the safety and health training before the employee starts a new job. In addition, safety and health training for new employees should include general and departmental safety and health rules.

WHO CONDUCTS SAFETY TRAINING PROGRAMS

Just who provides instruction for specific topics depends largely on the subjects to be taught, who is available, who is qualified, who has the ability to make effective presentations, and so on. With these factors considered, choose from among the following persons to provide various types of safety training programs.

- Director of training.
- Training department personnel.
- Director of safety.
- Plant engineer
- Safety engineer.
- Industrial engineer.
- First-line supervisors.
- Consultants.
- Insurance company safety specialists.
- Fire department personnel.
- Medical personnel.
- Machine, vehicle, and equipment operators.

HOW TO TIME SAFETY TRAINING

Safety training should be provided upon employment and thereafter, as needed. Prior to the assumption of specific employment responsibilities, orientation and on-the-job training should be scheduled. Once an employee has been assigned, however, training may be scheduled before, during, or after regularly scheduled working hours. Factors that must be carefully considered in the scheduling of safety training include union agreements, employee preferences, and operational needs as defined by managers and supervisors.

Other timing considerations include the assignment of new or additional responsibilities. It is unfair to both the employee and the company to assign other responsibilities to an employee before the appropriate safety training has been provided.

LOCATION OF SAFETY TRAINING PROGRAMS

When planning safety training programs, their locations must be considered. Depending on the type of training to be scheduled, a broad variety of locations are available.

- Workplace.
- Simulated workplace (trainer).
- Office.
- Classroom.
- Laboratory.
- Outside company premises.

Precisely which location(s) are selected for use depends on the type of training, the availability of certain locations, the willingness of managers and supervisors to release operational sites otherwise allocated to production activities, the suitability of the locations to the presentation of training, and similar factors.

PRESENTATION OF SAFETY TRAINING PROGRAMS

Every instructor, whether neophyte or old timer, has specific responsibilities associated with his or her presentation of safety training programs. These are as follows:

- **Size of Class.** A class should be limited to 15 to 25 students. If the number in the group to be instructed would exceed 25, several groups should be formed.
- **Meeting Place.** The room should be well ventilated, well lighted, and away from noise and other distractions. The room should be clean, orderly, and attractive.
- **Equipment.** The meeting place should be equipped with a chalkboard, chalk, erasers, and a pointer. It should have tables or other writing space for each student. Seats should be comfortable and arranged so that all the students can easily see and hear. A "U"-shaped seating arrangement, with the instructor at the open end of the "U," is sometimes ideal. Auditorium seating may also be used. The room should be provided with a screen for showing motion pictures, slides, or other visual aids.
- **The Instructor's Job.** The most important assets of an instructor are sincerity, enthusiasm, and a knowledge of the subject. It is not necessary to have a teaching or training background. A sincere desire to help others and enthusiasm about teaching are basic to effective instruction. The procedures outlined in this book will help the instructor present the material clearly and concisely. See Table 2-1.

The assignment should be given in advance so that the students will have read the material and can discuss the contents during the class period. However this may not be possible for the beginning session.

Table 2-1 Checklist for Instructors[a]

DIRECTIONS: This checklist serves as a guide to improve teaching methods. A good instructor should be able to answer *Yes* to at least eighteen of the questions. Under fifteen would be below average.

	Yes	No
1. Do you check your classroom for ventilation, lighting, seating arrangement, etc.?	___	___
2. Do you prepare a lesson plan?	___	___
3. Do you use communication aids whenever possible?	___	___
4. Do you preview all films and slides before showing?	___	___
5. Do you use test results to find weak points in your teaching?	___	___
6. Do you vary your type of presentation according to the material?	___	___
7. Do you stay on the subject?	___	___
8. Do you cover the material in each lesson?	___	___
9. Do you make any attempt to know your students and to learn their names?	___	___
10. Do you introduce yourself to each new class?	___	___
11. Do you make the objectives of the lesison clear?	___	___
12. Do you summarize each lesson?	___	___
13. Do you introduce each new subject and explain its importance?	___	___
14. Do you refrain from using sarcasm in your class?	___	___
15. Do you start your class on time and do you dismiss the class on time?	___	___
16. Do you have all your equipment ready and tested at class time?	___	___
17. Do you talk directly to the class and avoid such practices as staring at the floor, pacing the floor, and juggling a piece of chalk?	___	___
18. Do you keep your meeting place orderly?	___	___
19. Do you use words that are easily understood by the class?	___	___
20. Do you make assignments, and are they clear?	___	___

[a]Taken from *Instructors Guide for Use with Supervisors Safety Manual* Copyright by National Safety Council, Chicago. Used with permission.

In training program discussions the leader gets others to think and to talk. The best tools for promoting discussion are specific topics and provocative questions. For this reason, the prepared meeting plans include a number of topics and questions. After a discussion is started, the leader guides it by additional questions, but not the kind that can be answered "yes" or "no." Participation should be obtained from the entire group. One person should not be allowed to do all the talking.

Skill in leading a discussion is acquired with practice. Every time an instructor leads a discussion, he or she has an opportunity to improve.

- **Use of Visual Aids.** It is highly recommended that visual aids such as slides be previewed before showing to the class. In addition to the slides, aids such as personal protective equipment, guards, charts, and models may be used to good advantage.

- **Use of Tests.** Tests help the instructor emphasize the key points of a lesson. They also help the instructor in determining how good a job he or she is doing.

- **Use of Discussion Topics.** Discussion topics should be listed for each chapter emphasized in the class discussion. Since there may be others that apply to a specific operation or industry, the discussions should not be restricted to the topics listed here. It is a good idea to write on the chalkboard the points agreed on by the group. This will help to remind everyone of what has been discussed and to make sure that no important point is overlooked. The instrucotr is in the best position to determine the points that need stressing.

- **Lesson Plans.** Effective teaching calls for good planning. "Lesson plans," which should be written, standardize training. They also help the instructor to accomplish the following:
  - Present material in the proper order.
  - Avoid omission of essential material.
  - Conduct the sessions according to a timetable.
  - Place proper emphasis on items to be covered.
  - Provide for student participation.
  - Gain confidence.

For lesson plans to be really effective, each instructor should carefully plan the presentation. Table 2-2 gives a sample lesson plan. This is merely a suggested model to help in the careful planning of each lesson.

Table 2-2 Sample Lesson Plan: "Safety and the Supervisor"[a]

OBJECTIVES

To make supervisors aware that safety and production go together and that safety is their responsibility.

TRAINING AIDS AND MATERIALS

Distribute copies of the workbook and assign readings before the first meeting. Tell students to be prepared to discuss the reading assignments in class. Make sure that chalkboard, chalk, eraser, and pointer are available.

PRESENTATION

If you are not known to the group, write your name on the chalkboard and introduce yourself.

In this first session we want to discuss the problems of safety and production. Safety and production cannot be separated. Better production results from safety. We also want to emphasize the duties and responsibilities of a foreman.

It is well known that accidents cost money . . . but in many cases we see only the direct costs—medical expenses and compensation costs. We often fail to take into account the damage to equipment, lost production, and the cost of training replacements for injured workers.

(If possible, take a particular case from your own files and discuss it.)

To start this discussion, let's see if we can make a list of the duties of a first-line supervisor. We'll list the duties here on the board. (The list should include the following points: getting out production, maintaining orderly conditions, maintaining equipment, establishing work methods and procedures, supervising work, instructing workers, and maintaining worker morale.)

How many of these duties are connected with safety? (Ask members of the group to name details of their duties. Point out that maintaining orderly conditions, instructing workers, establishing work procedures, and maintaining morale are really methods of preventing accidents.)

Even without a safety program, how many of these duties would a first-line supervisor have to perform? Have the group name the duties for you to check, and point out that most of these are normal supervisory functions, which would be part of the attendees' jobs even if there were no formal safety program in the plant.

Can anyone but the first-line supervisor do these jobs regularly? Let's check over this list. Tell me who else could handle these jobs. I'll write his or her name or title after the item. (Allow students to discuss fully. Be sure they understand that no one else can perform these duties.)

Table 2-2 (*Continued*)

SUMMARY

From our discussion, it can readily be seen that safety and production cannot be separated. From your own listing, it is obvious that safety is a supervisory responsibility.

Even if no formal safety program existed, these duties would still have to be performed. It is only logical to conclude that if you are doing your job, you'll be preventing accidents. In preventing accidents, you'll have more time for planning and improving production.

[a] Taken from *Instructors Guide for Use with Safety Supervisors Manual* Copyright by National Safety Council, Chicago. Used with permission.

HOW TO MOTIVATE FOR SAFETY

Today the basic needs or wants that motivate people to be safety conscious fall into six clearly defined categories. A training instructor should try to have the class come up with them. Even if different names are used, it is usually easy to restate the categories as follows: self-preservation, service to others, satisfaction, security, social contact, and status.

- **Appeal to Self-Preservation.** The strongest motive we all possess is self-preservation. For some people it is the only motive to which a supervisor can appeal. The instructor should ask the group how the self-preservation instinct in people is awakened. For example, the following may occur:
    - Workers may suffer physically or be incapacitated if they have accidents.
    - They may face loss of income.
    - There may be anguish to the family.
- **Encourage Service to Others.** The person who jumps into the water to save a drowning child and the soldier who goes into the jungle to bring back a wounded friend voluntarily risk their lives for their fellow humans. We must make employees conscious of the effect of their bad safety habits on others at work and at home. Supervisors should be asked how they handle this phase of safety motivation.
- **Create Satisfaction.** Job satisfaction and safety go hand in hand. An exhaustive study by a medical doctor shows the following:

- Departments where there is a greater probability for the typical employee to be promoted have fewer accidents than those where jobs all seem to "dead end."
- Departments with the best suggestion records (particularly those where people receive rewards for their suggestions) tend to have fewer accidents.
- Accidents are more frequent in jobs of lower rated prestige.
- Supervisors should be asked what they can do, so employees get the kind of satisfaction from their jobs that will motivate them to work safely.

- **Provide Security.** A sense of security about one's job makes for safety. Researchers have found that during a recession, accidents among experienced persons increase even among people who previously had good safety records.

  More than half of all accidents take place when an employee is worried, apprehensive, or in some other low emotional state.

  Some elements of job security are beyond the control of the supervisor who cannot, for example, prevent layoffs. Often, however, insecurity stems from poor communication.

  Supervisors should be asked to list ways in which good communication can minimize insecurity.

- **Stimulate Social Contact.** Group loyalty plays a vital role in safety. Work groups on the job exert a heavy influence on safety attitudes. Strauss and Sayles suggest that safety is to a considerable extent a function of group attitudes. Further evidence suggests that risk-taking is largely a group phenomenon. In a situation where there is heavy pressure for production or where fellow employees encourage one another to take chances, almost everyone begins to take the hazardous shortcuts that lead to injury.*

- **Build Up Status.** Trace the relationship between positive safety attitudes and status. Ask a group to consider the following questions:
  - What is the status or standing of the safe worker in my department?
  - How do employees regard the person who submits a safety suggestion?
  - What is the status of the accident "repeater" in the eyes of his or her co-workers?

---

*George Strauss and Leonard R. Sayles, *PERSONNEL: The Human Problems of Management;* © 1972, pp. 187. Adapted by permission of Prentice-Hall, Inc., Englewood Cliffs, New Jersey.

INITIATING SELF-APPRAISAL

One psychologist claims that 50% of all accidents are due to improper attitudes on the part of workers, 30% to lack of job knowledge or skill, and 20% to bodily defects such as poor vision. Another psychologist states that personality factors are responsible for most accidents.

- **Change Attitudes.** Ask training classes to list attitudes that in their estimation affect safety. The list might include recklessness, carelessness, hostility, inattention, indifference, laziness, and arrogance. Develop these one by one, as in the example that follows:

    Why are people reckless? Most of the time it is because they seek attention. If the person who *dares* to break a rule is injured, he or she gets attention through sympathy and extra care; or, if the person is not injured, a supervisor warns about the seriousness of the offense. Meanwhile the people who are working safely are ignored by their fellow workers and their supervisors. Have the group offer suggestions on how a more constructive form of attention giving should be devised.

- **Overcome Poor Habits.** Ask the group to suggest how poor habits can be overcome. Ideas may include the following:
    - Don't preach. Instruct people properly so they won't learn wrong habits in the first place.
    - Change habits one by one. A person's behavior is made up of a number of habits. To try to have someone give up all habits at once leads to frustration. A person can master new habits only one at a time.
    - Permit no exceptions. A single step backward may go far to restore the old habit. Such phrases as "everyone does it," "conditions are different this time," and "only this once" all result in retrogressions.
    - Be patient. Time is needed to unlearn the old habit as well as to learn the new one.

3

Instructional Sequence for Job Training

NEW EMPLOYEE ORIENTATION

Safety orientation of the new employee is the job of everyone, from the supervisor, who sees to orderly indoctrination, to fellow workers who look after him or her during the first days on the job. The orientation procedure should have two objectives.

1. Employee protection: to impart the factual information necessary to the individual's safety.
2. Employee involvement: to make him or her an active member of the safety team.

Every effort should be made to develop a safety-oriented state of mind in each employee early in employment. Specifically, the orientation program should include at least the following topics.

- Company safety philosophy.
- Plant safety program.
- Pertinent safety rules.
- Protective equipment needs.

Job Safety Analysis

- Injury reporting and medical treatment.
- Location and use of emergency equipment.
- Employee participation in plant safety program.

JOB SAFETY ANALYSIS

Job safety analysis sheets (Table 3-1) should be prepared in advance of meeting with each group. At the end of the discussion, each employee should work out the sheet for each task under his or her supervision.

Table 3-1 Job Safety Analysis

Job Description: *Three Spindle Drill Press—Impeller 8406*
Job Location: *Bldg. 19-2 Pump Section*

Key Job Steps	Tool Used	Potential Health and Injury Hazard	Safe Practices, Apparel, and Equipment
Get material for operation	Tote box	Dropping tote box on foot	Wear safety shoes. Have firm grip on box.
		Back strain from lifting	Stress proper lifting methods.
		Picking up overloaded boxes.	Tell employee to get help or lighten load.
Inspect and set up drill press	Drill press	Check for defective machines	Do not operate if defective. Attach red or yellow "do not operate" tag.
		Chuck wrench not removed	Always remove chuck wrench immediately after use.
Drilling		Making adjustments when machine is running	Always stop spindle before making adjustments.
		Hair, clothing, or jewelry catching on spindle	Wear head covering, snug-fitting clothing. No loose sleeves. Avoid wearing rings, bracelets, or wrist watches.
		Spinning work or fixture	Use proper blocks or clamps to hold work and fixture securely.

Table 3-1 (*Continued*)

Key Job Steps	Tool Used	Potential Health and Injury Hazard	Safe Practices, Apparel, and Equipment
		Injury to hands—cuts, etc.	*Never wear gloves.* Use hook, brush, or other tool to remove chips. Use compressed air only when instructed.
		Drill sticks in work	Stop spindle, free drill by hand.
		Flying chips	Wear proper eye protection
		Pinch points at belts	Always stop press before adjusting belts.
		Broken drills	Do not attempt to force drill, apply pressure.

Signature_____
Date_____
Page____

Source: Joanne T. Widner (Editor), *Selected Readings in Safety,* Houston: International Safety Academy, 1973, p. 51.

Job safety analysis is a procedure that identifies the hazards or potential accidents associated with each step of a job. Additionally, it helps to develop solutions that will either eliminate or guard against hazards.

Discuss in detail with the group each of the following four basic steps in making a job safety analysis:

- Select the job to be analyzed.
- Break down the job into successeive steps.
- Identify the hazards and potential accidents.
- Develop ways to eliminate hazards and potential accidents.

Near the end of the meeting request that each participant select a task that he or she supervises and make up a job safety analysis on the sheets provided.

EMPLOYEE ANALYSIS

Assess Mental and Physical Characteristics

List on the board suggestions from the group of mental and physical characteristics that may have an adverse effect on safety. These should include the following:

Poor eyesight	Contagious disease
Muscular weakness	Defective hearing
Poor coordination	Lack of dexterity
Poor health	Slow mental reaction
Limited range of motion	Lack of emotional stability
Poor attitude	Epilepsy
Mental aberration	

Evaluate Knowledge and Skill

The group should understand what knowledge and skill factors affect safety. Draw from the group such factors as ignorance of correct methods, faulty work habits, and insufficient experience.

TRAINING FOR SAFETY

How many times have you heard a fellow supervisor say: "I just don't understand how this accident happened. I've told these people again and again how to do that operation safely. They seem to understand and do it right for a while, but then they go back to their old dangerous habits. Why won't they learn?"

It isn't always obstinacy; people just forget. It's all a part of the complex human personality. There are so many contradictory forces and influences to cope with, it's easy to slip unconsciously into old, familiar—and often dangerous—ways of doing things.

Many industrial accidents occur because the worker has not been trained adequately in the proper procedures for the job. The supervisor may have assumed that the worker knew how to handle the equipment or assignments—and failed to ask questions to determine how the task was being done. Observe these three rules and you'll avoid the complacency trap:

Rule #1. Never assume that employees know how to carry out a new assignment. This applies to the most experienced help as well as newcomers. Be prepared to show them exactly what is to be done and how to handle the necessary tools or machinery. It's true that it is often difficult to determine whether people understand an assignment because they are afraid they'll appear stupid if they ask questions. However they must be convinced that it is actually smarter to ask, especially if any danger is attached to the job.

Rule #2. Never assume that safety rules are remembered. Just because the instructor knows them thoroughly is no assurance that everyone else does. Avoid the common fallacy of "We've said it so often, everybody knows it by now."

Rule #3. In addition to making certain that people understand their jobs, frequent reminders should be given concerning overall safe procedures. If a department has a hazard peculiar to its type of work, the dangerous conditions should be pointed out often. The TV and radio advertisers know the value of repetition, even when it's irritating. People remember what they hear over and over again.

HOW TO PREPARE THE LEARNER

To gain the maximum possible advantage from a given instructional situation, it is important to place the potential trainee in an appropriate psychological state of mind. Receptivity is greatest when the trainee is ready to learn. The training area, the necessary equipment, the instructor, and so on, should all be in readiness at the appointed time.

Once the learner is in the training area, the instructor should be introduced. Self-introduction is appropriate for a small group, for a larger group a third party should do the honors.

Presentation of specific training topics should be preceded by an explanation of each and of its necessity for job safety. Identification of the direct relationship between accidents and costs should set the tone for the training that follows.

HOW TO DEMONSTRATE THE JOB

Full-time safety directors agree that successful safety and accident prevention programs depend largely on skilled and safety-conscious first-line supervisors who know how to train their workers continuously. Safety should be stressed right along with job skills when new workers receive instruction. The following proved steps of training should be applied:

1. Explain how to do the job safely.
2. Present the safe sequence of tasks.
3. Control the partial tryout.
4. Handle the complete tryout.
5. Answer questions properly, at the right time.
6. Follow-up techniques.

With the initial instructions for new workers completed, the safety training task is just beginning. A constant watch on the inexperienced workers is necessary to be sure they are following the rules and developing safe working habits.

TRAINING SEQUENCE CONTENT

Following the appropriate introductions and statements encouraging motivation, the instructor should explain the reasons for that particular training session.

Topics that must be included to complete each presentation are as follows:

Personal protective equipment.
Personnel positions and locations.
Personnel actions on the job.
Tools.
Equipment.
Special procedures.
Orderliness.

TRAINING PROGRAM CHARACTERISTICS

A quality safety instruction program should include the following elements:

- Safety demonstrations of all machines and equipment used in the plant.
- Close supervision of employees when they are operating equipment.

- Application of all safety procedures by a supervisor who sets an example by following regulations and safe practices zealously and cheerfully.
- Safety practices and procedures that are an integral part of all instruction.
- Safety regulations individually developed for each department, each machine, and each hazardous operation.
- An employee safety pledge to emphasize the individual's role in safety.
- Pledge cards, safety test results, anecdotal records, and similar information maintained in a personnel file system, to provide a constant source of evaluative material for assessing the effectiveness of instruction and as evidence of sincere intent in the event of an OSHA inspection.
- Safety tests, both written and performance, used as a measuring device designed to evaluate how well personnel have assimilated the information, habits, skills, and attitudes essential for their own safety and well-being.
- Routine shop safety inspections and written reports conducted regularly by supervisors and administrators. Administrative authorities informed of all weaknesses and needs in the program as soon as they are detected. (File copies of such reports maintained.)
- Under no circumstances should the supervisor use unsafe machines or areas; any hazards should be corrected or eliminated upon detection. Use of a hazardous item not permitted until safety requirements are met.
- A definite line of communication between supervisors and administrators to ensure that proper attention is given to all safety matters and recommendations from any source that will improve safety conditions.
- All safety rules and regulations enforced at all times.
- All safety recommendations made by administration carried out immediately to improve safety working conditions.
- Under no conditions should an instructor allow a class of untrained employees to work unsupervised with potentially dangerous tools or equipment.

TEACHING AIDS

A quality safety training program must take advantage of a multitude of teaching aids and visuals effectively incorporated into safety discussions and demonstrations where appropriate for the particular age and maturity of the class or for the type of tools or equipment. The National Safety Council, insurance companies, industries, and many manufacturers provide free or inexpensive items that can assist in the implementation of the safety training program. The following aids are examples of the types that can be included in the safety training program:

- **Posters.** Choose carefully and change regularly. "Wallpapering" the training area with posters only makes meaningless decoration.
- **Audiovisual Materials.** Films, filmstrips, slides, photographs, recordings, videotapes, and so on.
- **Tests and Quizzes.** As learning devices, as well as checks on understanding of safety instruction.
- **Instruction Sheets and Project Sheets.** Written procedural materials should always include pertinent safety precautions.
- **Suggestion Box.** Encourages employee-in-training participation in safety. Suggestions should be considered by the group-in-training, not by the trainer alone. Suitable recognition should be given for adopted suggestions.

4

OSHA Safety and Health Training Requirements

In response to criticism, OSHA has made available its special publication, OSHA 2254, "Training Requirements of OSHA Standards."

Because some employers are not completely familiar with all the OSHA standards that require them to provide certain types of training for their employees, OSHA 2254 has been assembled for use in determining the individual training needs of each company.

The training standards contained in OSHA 2254 are presented in numerical order from the Code of Federal Regulations (CFR), and the portion of the CFR that applies to each training standard is quoted. To facilitate the reader's familiarization with OSHA training requirements, pages 9 to 31 of OSHA 2254 are reprinted below.

Part 1910—Safety and Health Training Requirements for General Industry

The following occupational safety and health requirements, that relate to the training of employees, are commonly referred to as "general industry" standards. These training requirements have been extracted from Part 1910 of the *Code of Federal Regulations* (CFR) to stress the training programs required by employers. They cover the following subparts:
  G—Occupational Health and Environmental Control
  H—Hazardous Materials
  I—Personal Protective Equipment
  J—General Environmental Controls
  K—Medical and First Aid
  L—Fire Protection
  N—Materials Handling and Storage
  O—Machine and Machine Guarding
  Q—Welding, Cutting and Brazing
  R—Special Industries

Part 1910
Occupational Safety and Health Requirements for General Industry

CFR Part	Standard
1910.94(d)(9)(i) and (vi) **VENTILATION**	(9) *Personal protection.* (i) All employees working in and around open-surface tank operations must be instructed as to the hazards of their respective jobs, and in the personal protection and first aid procedures applicable to these hazards.

9

OSHA Safety and Health Training Requirements

(vi) . . . Respirators shall be approved by the U.S. Bureau of Mines, U.S. Department of the Interior (see 30 CFR Part II) and shall be selected by a competent industrial hygienist or other technically qualified source. Respirators shall be used in accordance with 1910.134(a), (b), and (c), and persons who may require them shall be trained in their use.

1910.94(d)(11)(v)

(11) *Inspection, maintenance, and installation.* (v) If, in emergencies, such as rescue work, it is necessary to enter a tank which may contain a hazardous atmosphere, suitable respirators, such as self-contained breathing apparatus: hose mask with blower, if there is a possibility of oxygen deficiency; or a gas mask, selected and operated in accordance with subparagraph (9)(vi) of this paragraph, shall be used. If a contaminant in the tank can cause dermatitis, or be absorbed through the skin, the employee entering the tank shall also wear protective clothing. At least one trained standby employee with suitable respirator, shall be present in the nearest uncontaminated area. The standby employee must be able to communicate with the employee in the tank and be able to haul him out of the tank with a lifeline if necessary.

1910.96(f)(3)(viii)
IONIZING RADIATION

(f) *Immediate evacuation warning signal.* (viii) All employees whose work may necessitate their presence in an area covered by the signal shall be made familiar with the actual sound of the signal—preferably as it sounds at their work location. Before placing the system into operation, all employees normally working in the area shall be made acquainted with the signal by actual demonstration at their work locations.

1910.106(b)(5)(vi)(v)(3)

(vi) *Flood areas.* (3) That station operators and other employees depended

10

FLAMMABLE AND COMBUSTIBLE LIQUIDS

1910.109(d)(3)(i)
EXPLOSIVES AND BLASTING AGENTS

1910.109(d)(3)(iii)

1910.109(g)(3)(iii)(a)

1910.111(b)(13)(ii)
STORAGE AND HANDLING OF ANHYDROUS AMMONIA

upon to carry out such instructions are thoroughly informed as to the location and operation of such valves and other equipment necessary to effect these requirements.

(3) *Operation of transportation vehicles.* (i) Vehicles transporting explosives shall only be driven by and be in the charge of a driver who is physically fit, careful, capable, reliable, able to read and write the English language, and not addicted to the use, or under the influence of intoxicants, narcotics, or other dangerous drugs, and not less than 21 years of age. He shall be familiar with the traffic regulations, State laws, and the provisions of this section.

(iii) Every motor vehicle transporting any quantity of Class A or Class B explosives shall, at all times, be attended by a driver or other attendant of the motor carrier. This attendant shall have been made aware of the class of the explosive material in the vehicle and of its inherent dangers, and shall have been instructed in the measures and procedures to be followed in order to protect the public from those dangers. He shall have been made familiar with the vehicle he is assigned, and shall be trained, supplied with the necessary means, and authorized to move the vehicle when required.

(3) *Bulk delivery and mixing vehicles.* (a) The operator shall be trained in the safe operation of the vehicle together with its mixing, conveying, and related equipment. The employer shall assure that the operator is familiar with the commodities being delivered and the general procedure for handling emergency situations.

(13) *Tank car unloading points and operations.* (ii) The employer shall insure that unloading operations are performed by reliable persons properly instructed and given the authority to monitor careful

11

1910.134(a)(3)
RESPIRATORY PROTECTION

1910.134(b)(1), (2) and (3)

compliance with all applicable procedures.

(a) *Permissible practice.* (3) The employee shall use the provided respiratory protection in accordance with instructions and training received.

(b) *Requirements for a minimal acceptable program.* (1) Written standard operating procedures governing the selection and use of respirators shall be established.

(2) Respirators shall be selected on the basis of hazards to which the worker is exposed.

(3) The user shall be instructed and trained in the proper use of respirators and their limitations.

1910.134(e)(2), (3) and (5)

(e) *Use of respirators.* (2) The correct respirator shall be specified for each job. The respirator type is usually specified in the work procedures by a qualified individual supervising the respiratory protective program. The individual issuing them shall be adequately instructed to insure that the correct respirator is issued.

(3) Written procedures shall be prepared covering safe use of respirators in dangerous atmospheres that might be encountered in normal operations or in emergencies. Personnel shall be familiar with these procedures and the available respirators.

(5) For safe use of any respirator, it is essential that the user be properly instructed in its selection, use, and maintenance. Both supervisors and workers shall be so instructed by competent persons. Training shall provide the men an opportunity to handle the respirator, have it fitted properly, test its face-piece-to-face seal, wear it in normal air for a long familiarity period, and finally, to wear it in a test atmosphere.

1910.134(e)(5)(i)

(i) Every respirator wearer shall receive

OSHA Safety and Health Training Requirements 47

1910.145(c)(1) (ii),(2)(ii) and (3)
SPECIFICATIONS FOR ACCIDENT PREVENTION SIGNS AND TAGS

1910.151(a) and (b)
MEDICAL SERVICES AND FIRST AID

1910.161(a)(2)
CARBON DIOXIDE EXTINGUISHING SYSTEMS

fitting instructions including demonstrations and practice in how the respirator should be worn, how to adjust it, and how to determine if it fits properly. To assure proper protection, the facepiece fit shall be checked by the wearer each time he puts on the respirator. This may be done by following the manufacturer's facepiece fitting instructions.

(1) *Danger signs.* (ii) All employees shall be instructed that danger signs indicate immediate danger and that special precautions are necessary.

(2) *Caution signs.* (ii) All employees shall be instructed that caution signs indicate a possible hazard against which proper precautions should be taken.

(3) *Safety instruction signs.* Safety instruction signs shall be used where there is a need for general instructions and suggestions relative to safety measures.

(a) The employer shall insure the ready availability of medical personnel for advice and consultation matters of plant health.

(b) In the absence of an infirmary, clinic, or hospital in near proximity to the workplace which is used for the treatment of all injured employees, a person or persons shall be adequately trained to render first aid. First aid supplies approved by the consulting physician shall be readily available.

(a) *General requirements.* (2) *Safety requirements.* In any use of carbon dioxide where there is a possibility that employees may be trapped in, or enter into atmospheres made hazardous by a carbon dioxide discharge, suitable safeguards shall be provided to insure prompt evacuation of and to prevent entry into such atmospheres and also to provide means for prompt rescue of any trapped personnel. Such safety items as personnel training,

1910.178(l)
POWERED INDUSTRIAL TRUCKS

1910.179(m)(3) (ix)
OVERHEAD AND GANTRY CRANES

1910.180(h)(3)(xii)
CRAWLER LOCOMOTIVE AND TRUCK CRANES

1910.217(e)(3)
MECHANICAL POWER PRESSES

1910.217(f)(2)

1910.218(a)(2) (i) thru (iv)
FORGING MACHINES

warning signs, discharge alarms, predischarge alarms and breathing apparatus shall be considered.

(l) *Operator training.* Only trained and authorized operators shall be permitted to operate a powered industrial truck. Methods shall be devised to train operators in the safe operation of powered industrial trucks.

(3) *Moving the load.* (ix) When two or more cranes are used to lift a load one qualified responsible person shall be in charge of the operation. He shall analyze the operation and instruct all personnel involved in the proper positioning, rigging of the load, and the movements to be made.

(3) *Moving the load.* (xii) When two or more cranes are used to lift one load, one designated person shall be responsible for the operation. He shall be required to analyze the operation and instruct all personnel involved in the proper positioning, rigging of the load, and the movements to be made.

(3) *Training of maintenance personnel.* It shall be the responsibility of the employer to insure the original and continuing competence of personnel caring for, inspecting, and maintaining power presses.

(2) *Instruction to operators.* The employer shall train and instruct the operator in the safe method of work before starting work on any operation covered by this section. The employer shall insure by adequate supervision that correct operating procedures are being followed.

(2) *Inspection and maintenance.* It shall be the responsibility of the employer to maintain all forge shop equipment in a condition which will insure continued safe operation. This responsibility includes:

(i) Establishing periodic and regular maintenance safety checks and keeping

OSHA Safety and Health Training Requirements

records of these inspections.

(ii) Scheduling and recording inspection of guards and point of operation protection devices at frequent and regular intervals.

(iii) Training personnel for the proper inspection and maintenance of forging machinery and equipment.

(iv) All overhead parts shall be fastened or protected in such a manner that they will not fly off or fall in event of failure.

1910.252(b)(1)(iii)
WELDING, CUTTING, AND BRAZING

(iii) *Instruction.* Workmen designated to operate arc welding equipment shall have been properly instructed and qualified to operate such equipment as specified in subparagraph (4) of this paragraph.

1910.252(c)(1) (iii)

(iii) *Personnel.* Workmen designated to operate resistance welding equipment shall have been properly instructed and judged competent to operate such equipment.

1910.252(c)(6)

(6) *Maintenance.* Periodic inspection shall be made by qualified maintenance personnel, and records of the same maintained. The operator shall be instructed to report any equipment defects to his supervisor and the use of the equipment shall be discontinued until safety repairs have been completed.

1910.252(d)(2) (xii)(c)

(xii) *Management.* (c) Insist that cutters or welders and their supervisors are suitably trained in the safe operation of their equipment and the safe use of the process.

1910.264(d)(1)(v)
LAUNDRY MACHINERY AND OPERATIONS

(v) *Instruction of employees.* Employees shall be properly instructed as to the hazards of their work and be instructed in safe practices, by bulletins, printed rules, and verbal instructions.

1910.266(c)(5) (i) thru (xi)
PULPWOOD LOGGING

(5) *Chain saw operations.* (i) Chain saw operators shall be instructed to daily inspect the saws to assure that all handles and guards are in place and tight, and that all controls function properly, and that the muffler is operative.

15

(ii) Chain saw operators shall be instructed to follow manufacturer's instructions as to operation and adjustment.

(iii) Chain saw operators shall be instructed to fuel the saw only in safe areas and not under conditions conducive to fire such as near men smoking, hot engine, etc.

(iv) Chain saw operators shall be instructed to hold the saw with both hands during operation.

(v) Chain saw operators shall be instructed to start the saw at least 10 feet away from fueling area.

(vi) Chain saw operators shall be instructed to start the saw only on the ground or when otherwise firmly supported.

(vii) Chain saw operators shall be instructed to be certain of footing and to clear away brush which might interfere before starting to cut.

(viii) Chain saw operators shall be instructed not to use engine fuel for starting fires or as a cleaning solvent.

(ix) Chain saw operators shall be instructed to shut off the saw when carrying it for a distance greater than from tree to tree or in hazardous conditions such as slippery surfaces or heavy underbrush. The saw shall be at idle speed when carried short distances.

(x) Chain saw operators shall be instructed to carry the saw in a manner to prevent contact with the chain and muffler.

(xi) Chain saw operators shall be instructed not to use the saw to cut directly overhead or at a distance that would require the operator to relinquish a safe grip on the saw.

1910.266(c)(6) (i) thru (xxi)

(6) *Stationary and mobile equipment.* (i) Equipment operators shall be instructed as to the manufacturer's recommendations for equipment operation, maintenance,

safe practices, and site operating procedures.

(ii) Equipment shall be kept free of flammable material.

(iii) Equipment shall be kept free of any material which might contribute to slipping and falling.

(iv) Engine of equipment shall be shut down during fueling, servicing, and repairs except where operation is required for adjustment.

(v) Equipment shall be inspected for evidence of failure or incipient failure.

(vi) The equipment operator shall be instructed to walk completely around machine and assure that no obstacles or personnel are in the area before startup.

(vii) The equipment operator shall be instructed to start and operate equipment only from the operator's station or from safe area recommended by the manufacturer.

(viii) Seat belt shall be provided on mobile equipment.

(ix) The equipment operator shall be instructed to check all controls for proper function and response before starting working cycle.

(x) The equipment operator shall be instructed to ground or secure all movable elements when not in use.

(xi) The equipment operator shall be advised of the load capacity and operating speed of the equipment.

(xii) The equipment operator shall be advised of the stability limitations of the equipment.

(xiii) The equipment operator shall be instructed to maintain adequate distance from other equipment and personnel.

(xiv) Where signalmen are used, the equipment operator shall be instructed to operate the equipment only on signal from the designated signalman and only when signal is distinct and clearly understood.

(xv) The equipment operator shall be instructed not to operate movable elements (boom, grapple, load, etc.) close to or over personnel.

(xvi) The equipment operator shall be instructed to signal his intention before operation when personnel are in or near the working area.

(xvii) The equipment operator shall be instructed to dismount and stand clear for all loading and unloading of his mobile vehicle by other mobile equipment. The dismounted operator shall be visible to loader operator.

(xviii) The equipment operator shall be instructed to operate equipment in a manner that will not place undue shock loads on wire rope.

(xix) The equipment operator shall be instructed not to permit riders or observers on machine unless approved seating and protection is provided.

(xx) The equipment operator shall be instructed to shut down the engine when the equipment is stopped, apply brake locks and ground elements before he dismounts.

(xxi) The equipment operator shall be instructed, when any equipment is transported from one job location to another, to transport it on a vehicle of sufficient rated capacity and the equipment shall be properly secured during transit.

1910.266(c)(7)

(7) *Explosives.* Only trained and experienced personnel shall handle or use explosives. Usage shall comply with the requirements of 1910.109.

1910.266(e)(1)
(iii) thru (vii)

(e) *Pulpwood harvesting.* (1) *Felling, general.*

(iii) Workers shall be instructed not to approach a feller closer than twice the height of the trees being felled until the

OSHA Safety and Health Training Requirements

	feller has acknowledged the signal of approach.
	(iv) Lodged trees shall be pulled to the ground at first opportunity with mechanical equipment or animal.
	(v) Workers shall be instructed not to work under a lodged tree.
	(vi) Special precautions shall be taken to prevent felling trees into powerlines.
	(vii) If a tree does make contact with a powerline, the power company shall be notified immediately and all personnel shall remain clear of the area until power company personnel advise that conditions are safe.
1910.266(e)(2) (i) and (ii)	(2) *Manual felling.* (i) The feller shall be instructed to plan retreat path and clear path as necessary before cut is started.
	(ii) The feller shall be instructed to appraise situation for dead limbs, the lean of tree to be cut, wind conditions, location of other trees and other hazards and exercise proper precautions before cut is started.
1910.266(e)(6) (viii)	(6) *Skidding and prehauling.* (viii) The operator shall be instructed to be observant and cautious of height of load and vehicle when traveling under trees, limbs, and other overhead obstructions.
1910.266(e)(9)	(9) *Off highway truck transport.* Truck drivers shall be instructed to stop their vehicles, dismount, check and tighten loose load binders, either just before or immediately after leaving a private road to enter a public road.
1910.1003 (e)(5) (i) thru (ii) **4-NITROBIPHENYL**	(5) *Training and indoctrination.* (i) Each employee prior to being authorized to enter a regulated area, shall receive a training and indoctrination program including but not necessarily limited to:
	(a) The nature of the carcinogenic hazards of 4-Nitrobiphenyl, including local and systemic toxicity;
	(b) The specific nature of the operation

involving 4-Nitrobiphenyl which could result in exposure;

(c) The purpose for and application of the medical surveillance program, including, as appropriate, methods of self-examination;

(d) The purpose for and application of decontamination practices and purposes;

(e) The purpose for and significance of emergency practices and procedures;

(f) The employee's specific role in emergency procedures;

(g) Specific information to aid the employee in recognition and evaluation of conditions and situations which may result in the release of 4-Nitrobiphenyl;

(h) The purpose for and application of specific first aid procedures and practices;

(i) A review of this section at the employee's first training and indoctrination program and annually thereafter.

(ii) Specific emergency procedures shall be prescribed, and posted, and employees shall be familiarized with their terms, and rehearsed in their application.

1910.1004 (e)(5)
(i) thru (ii)
ALPHA-NAPHTHYL-AMINE

(5) *Training and indoctrination.* (i) Each employee prior to being authorized to enter a regulated area, shall receive a training and indoctrination program including, but not necessarily limited to:

(a) The nature of the carcinogenic hazards of alpha-Naphthylamine, including local and systemic toxicity;

(b) The specific nature of the operation involving alpha-Naphthylamine which could result in exposure;

(c) The purpose for and application of the medical surveillance program, including, as appropriate, methods of self-examination;

(d) The purpose for and application of decontamination practices and purposes;

(e) The purpose for and significance of emergency practices and procedures;

(f) The employee's specific role in emergency procedures;

(g) Specific information to aid the employee in recognition and evaluation of conditions and situations which may result in the release of alpha-Naphthylamine;

(h) The purpose for and application of specific first aid procedures and practices;

(i) A review of this section at the employee's first training and indoctrination program and annually thereafter.

(ii) Specific emergency procedures shall be prescribed, and posted, and employees shall be familiarized with their terms, and rehearsed in their application.

1910.1005 (e)(5) (i) thru (ii)
4,4'-METHYLENE BIS(2-CHLORO-ANILINE)

(5) *Training and indoctrination.* (i) Each employee prior to being authorized to enter a regulated area, shall receive a training and indoctrination program including, but not necessarily limited to:

(a) The nature of the carcinogenic hazards of 4,4'-Methylene bis(2-chloroaniline), including local and systemic toxicity;

(b) The specific nature of the operation involving 4,4'-Methylene bis(2-chloroaniline) which could result in exposure;

(c) The purpose for and application of the medical surveillance program, including, as appropriate, methods of self-examination;

(d) The purpose for and application of decontamination practices and purposes;

(e) The purpose for and significance of emergency practices and procedures;

(f) The employee's specific role in emergency procedures;

(g) Specific information to aid the employee in recognition and evaluation of conditions and situations which may result in the release of 4,4'-Methylene bis(2-chloroaniline);

(h) The purpose for and application of specific first aid procedures and practices;

21

(i) A review of this section at the employee's first training and indoctrination program and annually thereafter.

(ii) Specific emergency procedures shall be prescribed, and posted, and employees shall be familiarized with their terms, and rehearsed in their application.

1910.1006 (e)(5) (i) thru (ii) **METHYL CHLORO-METHYL ETHER**

(5) *Training and indoctrination.* (i) Each employee prior to being authorized to enter a regulated area, shall receive a training and indoctrination program including, but not necessarily limited to:

(a) The nature of the carcinogenic hazards of Methyl chloromethyl ether, including local and systemic toxicity;

(b) The specific nature of the operation involving Methyl chloromethyl ether which could result in exposure;

(c) The purpose for and application of medical surveillance program, including, as appropriate, methods of self-examination;

(d) The purpose for and application of decontamination practices and purposes;

(e) The purpose for and significance of emergency practices and procedures;

(f) The employee's specific role in emergency procedures;

(g) Specific information to aid the employee in recognition and evaluation of conditions and situations which may result in the release of Methyl chloromethyl ether;

(h) The purpose for and application of specific first aid procedures and practices;

(i) A review of this section at the employee's first training and indoctrination program and annually thereafter.

(ii) Specific emergency procedures shall be prescribed, and posted, and employees shall be familiarized with their terms, and rehearsed in their application.

1910.1007 (e)(5) (i) thru (ii)

(5) *Training and indoctrination.* (i) Each employee prior to being authorized

3,3′-DICHLORO-BENZIDINE (or its salts)

to enter a regulated area, shall receive a training and indoctrination program including, but not necessarily limited to:

(a) The nature of the carcinogenic hazards of 3,3′-Dichlorobenzidine (or its salts), including local and systemic toxicity;

(b) The specific nature of the operation involving 3,3′-Dichlorobenzidine (or its salts) which could result in exposure;

(c) The purpose for and application of the medical surveillance program, including, as appropriate, methods of self-examination;

(d) The purpose for and application of decontamination practices and purposes;

(e) The purpose for and significance of emergency practices and procedures;

(f) The employee's specific role in emergency procedures;

(g) Specific information to aid the employee in recognition and evaluation of conditions and situations which may result in the release of 3,3′-Dichlorobenzidine (or its salts);

(h) The purpose for and application of specific first aid procedures and practices;

(i) A review of this section at the employee's first training and indoctrination program and annually thereafter.

(ii) Specific emergency procedures shall be prescribed, and posted, and employees shall be familiarized with their terms, and rehearsed in their application.

1910.1008 (e)(5) (i) thru (ii)
BIS-CHLOROMETHYL ETHER

(5) *Training and indoctrination.* (i) Each employee prior to being authorized to enter a regulated area, shall receive a training and indoctrination program including, but not necessarily limited to:

(a) The nature of the carcinogenic hazards of bis-Chloromethyl ether, including local and systemic toxicity;

(b) The specific nature of the operation involving bis-Chloromethyl ether which could result in exposure;

(c) The purpose for and application of the medical surveillance program, including, as appropriate, methods of self-examination;

(d) The purpose for and application of decontamination practices and purposes;

(e) The purpose for and significance of emergency practices and procedures;

(f) The employee's specific role in emergency procedures;

(g) Specific information to aid the employee in recognition and evaluation of conditions and situations which may result in the release of bis-Chloromethyl ether;

(h) The purpose for and application of specific first aid procedures and practices;

(i) A review of this section at the employee's first training and indoctrination program and annually thereafter.

(ii) Specific emergency procedures shall be prescribed, and posted, and employees shall be familiarized with their terms, and rehearsed in their application.

1910.1009 (e)(5)
(i) thru (ii)
BETA-NAPHTHYLAMINE

(5) *Training and indoctrination.* (i) Each employee prior to being authorized to enter a regulated area, shall receive a training and indoctrination program including, but not necessarily limited to:

(a) The nature of the carcinogenic hazards of beta-Naphthylamine, including local and systemic toxicity;

(b) The specific nature of the operation involving beta-Naphthylamine which could result in exposure;

(c) The purpose for and application of the medical surveillance program, including, as appropriate, methods of self-examination;

(d) The purpose for and application of decontamination practices and purposes;

(e) The purpose for and significance of emergency practices and procedures;

(f) The employee's specific role in

emergency procedures;

(g) Specific information to aid the employee in recognition and evaluation of conditions and situations which may result in the release of beta-Naphthylamine;

(h) The purpose for and application of specific first aid procedures and practices;

(i) A review of this section at the employee's first training and indoctrination program and annually thereafter.

(ii) Specific emergency procedures shall be prescribed, and posted, and employees shall be familiarized with their terms, and rehearsed in their application.

1910.1010 (e)(5)
(i) thru (ii)
BENZIDINE

(5) *Training and indoctrination.* (i) Each employee prior to being authorized to enter a regulated area, shall receive a training and indoctrination program including, but not necessarily limited to:

(a) The nature of the carcinogenic hazards of Benzidine, including local and systemic toxicity;

(b) The specific nature of the operation involving Benzidine which could result in exposure;

(c) The purpose for and application of the medical surveillance program, including, as appropriate, methods of self-examination;

(d) The purpose for and application of decontamination practices and purposes;

(e) The purpose for and significance of emergency practices and procedures;

(f) The employee's specific role in emergency procedures;

(g) Specific information to aid the employee in recognition and evaluation of conditions and situations which may result in the release of Benzidine;

(h) The purpose for and application of specific first aid procedures and practices;

(i) A review of this section at the employee's first training and indoctrination program and annually thereafter.

1910.1011 (e)(5)
(i) thru (ii)
4-AMINODIPHENYL

(ii) Specific emergency procedures shall be prescribed, and posted, and employees shall be familiarized with their terms, and rehearsed in their application.

(5) *Training and indoctrination.* (i) Each employee prior to being authorized to enter a regulated area, shall receive a training and indoctrination program including, but not necessarily limited to:

(a) The nature of the carcinogenic hazards of 4-Aminodiphenyl, including local and systemic toxicity;

(b) The specific nature of the operation involving 4-Aminodiphenyl which could result in exposure;

(c) The purpose for and application of the medical surveillance program, including, as appropriate, methods of self-examination;

(d) The purpose for and application of decontamination practices and purposes;

(e) The purpose for and significance of emergency practices and procedures;

(f) The employee's specific role in emergency procedures;

(g) Specific information to aid the employee in recognition and evaluation of conditions and situations which may result in the release of 4-Aminodiphenyl;

(h) The purpose for and application of specific first aid procedures and practices;

(i) A review of this section at the employee's first training and indoctrination program and annually thereafter.

(ii) Specific emergency procedures shall be prescribed, and posted, and employees shall be familiarized with their terms, and rehearsed in their application.

1910.1012 (e)(5)
(1) thru (ii)
ETHYLENEIMINE

(5) *Training and indoctrination.* (i) Each employee prior to being authorized to enter a regulated area, shall receive a training and indoctrination program including, but not necessarily limited to:

(a) The nature of the carcinogenic

hazards of Ethyleneimine, including local and systemic toxicity;

(b) The specific nature of the operation involving Ethyleneimine which could result in exposure;

(c) The purpose for and application of the medical surveillance program, including, as appropriate, methods of self-examination;

(d) The purpose for and application of decontamination practices and purposes;

(e) The purpose for and significance of emergency practices and procedures;

(f) The employee's specific role in emergency procedures;

(g) Specific information to aid the employee in recognition and evaluation of conditions and situations which may result in the release of Ethyleneimine;

(h) The purpose for and application of specific first aid procedures and practices;

(i) A review of this section at the employee's first training and indoctrination program and annually thereafter.

(ii) Specific emergency procedures shall be prescribed, and posted, and employees shall be familiarized with their terms, and rehearsed in their application.

1910.1013 (e)(5)
(i) thru (ii)
BETA-PROPIOLACTONE

(5) *Training and indoctrination.* (i) Each employee prior to being authorized to enter a regulated area, shall receive a training and indoctrination program including, but not necessarily limited to:

(a) The nature of the carcinogenic hazards of beta-Propiolactone, including local and systemic toxicity;

(b) The specific nature of the operation involving beta-Propiolactone which could result in exposure;

(c) The purpose for and application of the medical surveillance program, including, as appropriate, methods of self-examination;

(d) The purpose for and application of

decontamination practices and purposes;

(e) The purpose for and significance of emergency practices and procedures;

(f) The employee's specific role in emergency procedures;

(g) Specific information to aid the employee in recognition and evaluation of conditions and situations which may result in the release of beta-Propiolactone;

(h) The purpose for and application of specific first aid procedures and practices;

(i) A review of this section at the employee's first training and indoctrination program and annually thereafter.

(ii) Specific emergency procedures shall be prescribed, and posted, and employees shall be familiarized with their terms, and rehearsed in their application.

1910.1014 (e)(5)
(i) thru (ii)
2-ACETYLAMINO-FLUORENE

(5) *Training and indoctrination.* (i) Each employee prior to being authorized to enter a regulated area, shall receive a training and indoctrination program including, but not necessarily limited to:

(a) The nature of the carcinogenic hazards of 2-Acetylaminofluorene including local and systemic toxicity;

(b) The specific nature of the operation involving 2-Acetylaminofluorene which could result in exposure;

(c) The purpose for and application of medical surveillance program, including, as appropriate, methods of self-examination;

(d) The purpose for and application of decontamination practices and purposes;

(e) The purpose for and significance of emergency practices and procedures;

(f) The employee's specific role in emergency procedures;

(g) Specific information to aid the employee in recognition and evaluation of conditions and situations which may result in the release of 2-Acetylaminofluorene;

(h) The purpose for and application of

specific first aid procedures and practices;

(i) A review of this section at the employee's first training and indoctrination program and annually thereafter.

(ii) Specific emergency procedures shall be prescribed, and posted, and employees shall be familiarized with their terms, and rehearsed in their application.

1910.1015 (e)(5) (i) thru (ii)

4-DIMETHYLAMINOAZO-BENZENE

(5) *Training and indoctrination.* (i) Each employee prior to being authorized to enter a regulated area, shall receive a training and indoctrination program including, but not necessarily limited to:

(a) The nature of the carcinogenic hazards of 4-Dimethylaminoazobenzene, including local and systemic toxicity;

(b) The specific nature of the operation involving 4-Dimethylaminoazobenzene which could result in exposure;

(c) The purpose for and application of the medical surveillance program, including, as appropriate, methods of self-examination;

(d) The purpose for and application of decontamination practices and purposes;

(e) The purpose for and significance of emergency practices and procedures;

(f) The employee's specific role in emergency procedures;

(g) Specific information to aid the employee in recognition and evaluation of conditions and situations which may result in the release of 4-Dimethylaminoazobenzene;

(h) The purpose for and application of specific first aid procedures and practices;

(i) A review of this section at the employee's first training and indoctrination program and annually thereafter..

(ii) Specific emergency procedures shall be prescribed, and posted, and employees shall be familiarized with their terms, and rehearsed in their application.

1910.1016 (e)(5)
(i) thru (ii)
N-NITROSODI-METHYLAMINE

(5) *Training and indoctrination.* (i) Each employee prior to being authorized to enter a regulated area, shall receive a training and indoctrination program including, but not necessarily limited to:

(a) The nature of the carcinogenic hazards of N-Nitrosodimethylamine, including local and systemic toxicity;

(b) The specific nature of the operation involving N-Nitrosodimethylamine which could result in exposure;

(c) The purpose for and application of the medical surveillance program, including, as appropriate, methods of self-examination;

(d) The purpose for and application of decontamination practices and purposes;

(e) The purpose for and significance of emergency practices and procedures;

(f) The employee's specific role in emergency procedures;

(g) Specific information to aid the employee in recognition and evaluation of conditions and situations which may result in the release of N-Nitrosodimethylamine;

(h) The purpose for and application of specific first aid procedures and practices;

(i) A review of this section at the employee's first training and indoctrination program and annually thereafter.

(ii) Specific emergency procedures shall be prescribed, and posted, and employees shall be familiarized with their terms, and rehearsed in their application.

1910.1017 (j)(1)
(i) thru (ix)
VINYL CHLORIDE

(j) *Training.* Each employee engaged in vinyl chloride or polyvinyl chloride operations shall be provided training in a program relating to the hazards of vinyl chloride and precautions for its safe use.

(1) The program shall include:

(i) The nature of the health hazard from chronic exposure to vinyl chloride including specifically the carcinogenic hazard;

(ii) The specific nature of operations

which could result in exposure to vinyl chloride in excess of the permissible limit and necessary protective steps;

(iii) The purpose for, proper use, and limitations of respiratory protective devices;

(iv) The fire hazard and acute toxicity of vinyl chloride, and the necessary protective steps;

(v) The purpose for and a description of the monitoring program;

(vi) The purpose for, and a description of, the medical surveillance program;

(vii) Emergency procedures;

(viii) Specific information to aid the employee in recognition of conditions which may result in the release of vinyl chloride; and

(ix) A review of this standard at the employee's first training and indoctrination program, and annually thereafter.

5

The Supervisor's Role in Safety Training

Every member of management from plant manager to first-line supervisor is directly responsible for the safety of the employees under him or her. The first-line supervisor, however, is the key to a good safety record. Why? Because studies have shown that time and time again companies with good safety program organization still have many accidents. Why? Because even though the rules were written down, nobody told the employees WHY they should follow them. Also, many times employees were simply not told HOW to do their work safely. Thus the supervisor has responsibilities running from maintaining good safety morale to setting high standards of housekeeping; from selecting the best person for the job to providing adequate training for all employees; from ensuring proper recognition for jobs well done to imposing discipline where needed. The section "Provide a Safety Climate" contains a few rules of thumb for effective safety supervision.

PROVIDE SAFETY CLIMATE

Set the Example

A supervisor sets the pace and greatly influences the employee's interest in safety by his or her actions. The examples set by the supervisor should reflect sincerity and alertness. Some specific suggestions are given below.

Provide Safety Climate

- Observe all safety and fire protection rules. The supervisor who makes exceptions to such rules for personal convenience seriously undermines the safety effort in that area.
- Demonstrate sincere concern about safety by seriously and consistently "preaching and practicing" safety.
- Keep safety on at least an equal level of importance with quality and quantity of output.
- Ignore no violation of safety regulations.
- Wear personal protective equipment where required. It's a good way to "sell" it and to demonstrate that using such protection is the intelligent thing to do.
- Discuss some aspect of safety with employees every day.
- Be enthusiastic about safety and fire protection. The enthusiasm that the supervisor displays will generate enthusiasm in the employees. Give safety priority among your problems. Never let quality, production, or cost considerations compromise safety or essential fire protection.

Cooperate with Others

Outline the responsibility of the company's safety department or the functions of the safety engineer. If possible, have a representative address the meeting. Have group members cite ways in which they cooperate with the safety department.

Discuss: "If someone from the Safety Department told one of your employees to stop what he or she was doing because the person was not following safety rules, what would you do?" "If you saw an employee of *another supervisor* violate a safety rule, what would you do?" "If you saw another supervisor violate a safety rule in the presence of employees of your department, what would you do?"

Sell a "Series of Successes"

"Teaching takes place most effectively when people see that they are *successful*—when they experience a "series of successes." Have the group suggest ways of putting the "S.O.S." principle to work in safety training. Responses should include the following:

Make safety a part of job instruction.

Instruct not only the new employee in safe practices, but also the older employee who has been transferred to a new job.

Draft understandable safety rules.

Make sure instructions are easy to follow.

Always give reasons for safety instruction.

Have employees "feed back" instructions to you, so *you* know *they* know.

While coaching and correcting people, make a particular point of commenting whenever a job is done safely, a precaution is taken, or a safety rule is properly followed.

Give recognition for what is done correctly.

Reduce "Accident Proneness"

Some years ago a large company embarked on an intensive campaign to reduce off-the-job accidents. It was found that some employees were 15 times as safe on the job as off the job.

These findings led to the conclusion that environment, and good safety programs, are probably more important than whether an employee is "accident prone."

Call on the group for ways to make the work environment more conducive to safety. The participants should suggest the following measures:

Keep all safety devices in good order.

Be sure the safety device does not slow up the work.

Sell the worker on the value of the safety device.

Never permit operation without the device in place and in proper use.

Vary work assignments to avoid monotony.

Keep the workplace as free of distractions as possible.

Use posters as reminders on how to follow safe practices.

Study Safety Design

Frederick W. Taylor, the "father of scientific management," was the first to call attention to the fact that people are an important factor in the work situation. His findings led to increased productivity. He pointed out that just as there was a best machine for a job, so were there best ways for people to do their jobs. His principles of motion study were directed to *adapting people to machines.*

Today design engineers are adapting machines to people. They study what motions are most natural to employees, and what reactions they have to noise and other stimuli, then they design machinery accordingly. We might say that machines are "tailored" to people.

Not only does this concern for people affect productivity, but safety is improved as well.

SUPERVISORY ACTIONS

To appreciate and evaluate fully the safety and fire hazards involved on a given job, the supervisor should understand thoroughly the entire process or operation for which he or she is responsible.

Every trip through the plant should be an impromptu inspection tour. In this way the supervisor can correct hazards that might otherwise cause injuries.

Safe working conditions can be achieved only through the detection and elimination of unsafe conditions and unsafe practices. Inspections help to do this. Some inspections can and should be made by the safety section and inspection committees, but there is no substitute for a firsthand look by the supervisor. Where practical, including wage roll people on inspection teams is a good way to show them that they have a part in the safety effort. It also gives the supervisor a chance to illustrate the standards of performance he is seeking.

To be of value, the supervisor's observations in the field must be translated into effective corrective action. It should be made clear that correction of an unsafe practice is not a reprimand in itself, but a step toward improved safety performance. Correction must be prompt to be effective.

When reasonable safety standards are not met and there are no extenuating circumstances, disciplinary action may be in order. It should be as consistent and equitable as possible, to keep employee resentment to a minimum. The improvement of performance and safety awareness is the object.

The ability of an employee to perform a specific job is dependent on his or her education, training, experience, and general capabilities. To achieve the safest, most efficient performance, the supervisor must know these characteristics when planning job assignments, training programs, and performance reviews.

SPECIAL PROCEDURES

Every company is unique and will no doubt need to establish some special safety-related procedures. Among the common ones that should be developed fairly quickly are the following: what to do regarding an employee request for equipment and/or facility shutdown when the employee believes an imminent danger exists, plans for emergency evacuation for fire or bad weather, postaccident procedures and hazard reporting. A form can be used for reporting either building or equipment physical hazards or personnel hazards such as poor safety practices.

It should be noted that the standardizing of postaccident procedures is no easy task: first aid must be administered by a qualified employee, the injured party must be transported to adequate medical treatment, and a qualified physician must decide when a medically treated employee can safely return to the job or to some less demanding activity.

When these special procedures have been drafted, they should be brought before the employee and executive safety committees for review and comment. Once approved at the committee level, these proposed operating procedures should receive approval by the firm's top management.

It is always necessary to make sure that all employees who will be influenced by a new procedure are completely informed of its impact on them. Communications such as those described in the earlier discussion on the company safety policy should be used to convey the information to as many employees as possible, with maximum feasible speed and clarity.

To be certain that the information has been accurately transmitted and received, supervisors should conduct spot checks. These should not be considered as exams with a pass/fail conclusion, but rather as a communications check to verify that the contents of a new procedure are genuinely understood by the people who must implement it.

OTHER GOALS AND OBJECTIVES

There are a wide variety of important projects, some long term and some short term, which should be part of any good company safety program.

Every company should have a *safety manual* that is easily understood by an overwhelming majority of the employees, and each employee should have a copy. A good safety manual has plenty of pictures, even cartoons, and straight talk on how to do a job safely.

Every member of both safety committees and every person in the maintenance or physical plant department should be provided with a copy of OSHA 2201, the OSHA *Safety and Health Standards Digest for General Industry*. This pocket-sized manual, which concisely explains about 90% of the basic applicable standards, was compiled to support voluntary compliance with OSHA workplace standards. This booklet is easily obtained from the closest OSHA office.*

About once a year, an employee attitude survey regarding the company safety and health program should be conducted. This will enable the firm to appropriately direct its safety efforts based on the most current perceptions of the employees.

---

*The address and telephone number can be located in the white pages of any telephone book. Look under United States Government, Department of Labor, Occupational Safety and Health Administration.

Other Goals and Objectives 71

Nearly every organization has one or more internal publications. For the most part these newspapers or newsletters are widely read by most employees, primarily because the articles are about their friends and fellow employees. A regular column such as "Spotlight on Safety" is excellent for reinforcing the concept that the only right way to do the job is the safe way.

When a project gets beyond a firm's internal expertise, outside help should be solicited. Two examples of such complex technical projects are a *noise suppression and/or containment project* and a *behavior modification program*. The former requires extensive engineering expertise, which is frequently available only through outside specialists. The latter will certainly call for the skills of an experienced industrial psychologist, who will design a customized procedure whereby employees receive *immediate* rewards for safe behaviors (e.g., reporting hazards, unsafe acts performed by others, and safe acceptable acts performed by themselves).

Emergency information signs containing accurate up-to-date information should be conspicuously posted throughout the plant. These signs should provide information on emergency evacuation routes (which have already been approved by the local fire marshal), fire extinguishers, oxygen bottles, fire hoses, emergency exits, and so on. These signs, with the inclusion of "you are here" stickers to identify a sign reader's present location, can be critically important in the event of an emergency. Persons who regularly work in one building but are not familiar with certain other portions of the overall facility are potential users of these signs, as are new employees, visitors, and the like.

It is important to regularly review the plant and its equipment. This can be accomplished only through a continuing series of *inspections* conducted by a variety of people. No-notice inspections conducted by the director of safety, the safety engineer, and company managers, audit inspections conducted monthly by members of the executive safety committee, semiannual inspections conducted by insurance company safety engineers and by local fire marshals, as well as those conducted by state occupational safety and health department personnel, should help eliminate most plant and equipment hazards.

Training and use of *accident investigators* can result in determining the real reasons for accidents, not just such old clichés as "he should have used a ladder" and "she wasn't using her eye protection." By playing down the obvious fault-finding and playing up the hazard elimination feature of this program, the director of safety has an excellent additional tool to help get the job done.

When a company doesn't have a first aid facility, it is important to have members of the employee safety committee conduct regular content reviews of all first aid kits. Otherwise it may not be learned that important first aid materials are missing until after an accident has occurred. Where a master first aid facility has been made available, the room should be accessible only to personnel who hold current first aid cards. By the way, first aid cards are valid for only three years following certification. A master roster of personnel authorized to administer first aid (by department and by shift) should be maintained by the director of personnel. Since most people rarely have the opportunity to put their training into practice, annual updating is necessary to ensure that any injured employees will receive first aid only from personnel whose first aid training is current.

Another way of communicating with all employees about safety is to hold small group meetings. Use a system such as Safety Day/Part I—1981. Twice annually for 30 minutes for all employees should be the minimum exposure to the director of safety, who can make presentations on relevant topics.

Other projects that can help reinforce safety and safe work practices on a site consist of the usual safety bulletin boards, safety posters, safety films, and selected guests for safety committee meetings.

No single project is going to be a panacea. Through the introduction of many such projects in a genuine attempt to enhance plant safety, however, positive results can be expected.

MANAGEMENT'S OVERVIEW OF SAFETY TRAINING

An effective accident prevention program requires proper job performance from everyone in the workplace.

All employees must know about the materials and equipment they are working with, what hazards are known in the operation, and how these hazards have been controlled or eliminated.

Each individual employee needs to know and understand the following points (especially if they have been included in the company policy and in a "code of safe practices"):

> No employee is expected to undertake a job until he or she has received instruction on how to do it properly and has been authorized to perform that job.
>
> No employee should undertake a job that appears to be unsafe.
>
> Mechanical safeguards are in place and must be kept in place.
>
> Each employee is expected to report all unsafe conditions encountered during work.
>
> Even slight injury or illness suffered by an employee must be reported at once.
>
> The safety and health program can be amplified as needs and new methods are discovered.

In addition to the points above, any safety rules that are a condition of employment, such as the use of safety shoes or eye protection, should be explained clearly and enforced at once.

The first-line supervisors must know how to train employees in the proper way of doing their jobs. Encourage and consider providing for supervisory training for these supervisors. (Many colleges offer appropriate introductory management training courses.)

In addition, some specific training requirements in the OSHA standards must be met, such as those that pertain to first aid and powered industrial trucks (including forklifts). In general, they deal with situations where the use of untrained or improperly trained operators on skill machinery could cause hazardous situations to develop, not only for the operator, but possibly for nearby workers, too.

Particular attention must be given to new employees. Immediately on arriving at work, new employees begin to learn things and to form attitudes about the company, the job, their boss, and their fellow employees. Learning and attitude formation occur regardless of whether the employer makes a training effort. If the new employees are trained during those first few hours and days to do things the right way, considerable losses may be avoided later.

At the same time, attention must be paid to regular employees, including the old timers. Old habits can be wrong habits. An employee who continues to repeat an unsafe procedure is not working safely, even if an "accident" has not resulted from this behavior.

Although every employee's attitude should be one of determination that "accidents" can be prevented, one thing more may be needed. It should be stressed that the responsibility assigned to the person in charge of the job—as well as to all other supervisors—is to be sure that there is a concerted effort underway at all times to follow every safe work procedure and health practice applicable to that job. It should be clearly explained to these supervisors that they should never silently condone unsafe or unhealthful activity in or around any workplace.

Finally, here are some less specific signals that might indicate a need for training or retraining:

Excessive waste or scrap.

High labor turnover.

An increase in the number of near misses that could have resulted in accidents.

A recent upswing in actual "accident" experience.

High injury and illness incidence.

Expansion of business and/or new employment.

A change in process or a new process.

6

Training in Safety Recordkeeping

Why statistics? During the course of a year many incidents may occur throughout a company. Some may result in serious injuries, fires, or explosions. In others the severity is slight, but the potential for serious injury or major property loss is great. It is important for a company to evaluate these experiences so that it may gauge its progress in the field of safety and fire protection, and take steps to prevent recurrences.

CLASSIFYING ACCIDENTS

A company's classification of injuries is generally based on and is in accordance with ANSI Method of Recording and Measuring Work Injury Experience, ANSI Z16.1-1973, published by the American National Standards Institute, 1430 Broadway, New York City, New York 10018. This standard is sponsored by the National Safety Council and is generally accepted by American industry.

Occupational Injuries

Basically, an injury sustained in the course of employment is an occupational injury. The term "injury" includes occupational disease. Occupational injuries are classified by severity into three categories:

Disabling or Lost-Time Injury. Death, and causes resulting in any degree of permanent disability that interferes with normal physiological functions, and loss of time. "Loss of time" is defined as the inability of an employee to return to work at an established job on the day following the injury or any subsequent day, some time during the hours corresponding to the shift on which the person was injured, whether scheduled or not.

Medical Case. Those injuries requiring the services of a doctor, but which do not result in permanent disability or lost time. Injuries of this nature generally require that a "First Report of Injury" be filed with the appropriate state agency.

Minor Injury. Those cases requiring first aid treatment only, with no disability or interference with duties.

RECORDS MAINTENANCE

No company safety program is complete without records of what has been planned and achieved in the past, what is in process, and what is planned for the future. Complete documentation for a company's internal program review, as well as to satisfy outside inspectors, is an absolute necessity. See Table 6-1.

Some of the specific records and files that should be carefully maintained are OSHA form 200 (the five-year chronological log and annual summary) and OSHA form 101 (the equivalent of the workmen's compensation accident report). These two are musts as required by OSHA.

Other records and files that should be readily available are the Special Computerized Accident Report (SCAR), a monthly analysis of the company accident history by department and by shift. Printouts (SCAR tissue) include monthly, quarterly, and annual summaries of accident frequency and accident severity rates.

The Medical Alert File is a loose-leaf binder consisting of a separate page for each employee. If an employee is rendered unconscious and is in need of medical assistance, a physician may consult this file for such information as required or prohibited medicines.

Records of accident investigations, hazard inspections, audit inspections, no-notice inspections, insurance company and fire department inspections, as well as training programs, should be maintained. Other records that should be maintained include exposure of employees to toxic substances and sources, employee physical examinations, and employment records.

Every business has ideas on the best place and the best way to keep safety records. A master file maintained by the director of safety with subfiles main-

Table 6-1 Requirements for Safety Files

The following items are presented for your convenience as you review your administrative storage index to determine the adequacy of your safety-related files.

No.	Action Required	Action Completed Yes	No
1.	Is there a separate section for safety-related files?		
2.	Are the following subjects provided for in the safety section of the files:		
	a. Blank OSHA forms?		
	b. Completed OSHA forms?		
	c. Blank Company safety forms?		
	d. Completed company safety forms?		
	e. Blank safety checklists?		
	f. Completed safety checklists?		
	g. Agendas of company safety meetings?		
	h. Minutes of company safety meetings?		
	i. Records of safety equipment purchases?		
	j. Records of safety equipment checkouts?		
	k. Incoming correspondence related to safety?		
	l. Outgoing correspondence related to safety?		
	m. Record of safety projects assigned?		
	n. Record of safety projects completed?		
	o. Record of fire drills (if applicable)?		
	p. Record of external assistance used to provide specialized safety expertise?		
	q. Record of inspections by fire department, insurance companies, state and city inspectors, and OSHA compliance officers?		
	r. National Safety Council catalogs and brochures for films, posters, and other safety-related materials?		
3.	Are the files listed in item 2 reviewed periodically:		
	a. To insure that they are current?		
	b. To retire material over five years old?		
4.	Are the safety-related files reviewed periodically to determine the need to eliminate selected files and to add new subjects?		
5.	Is the index to the file current, so that an outsider could easily understand the system?		

tained by day-to-day users is recommended. For example, records of employee orientations, as well as the Medical Alert File and OSHA records, should be in the personnel office. Records of hazard corrections should be kept by the safety engineer, as should records of internally and externally sponsored inspections.

INJURY EXPERIENCE MEASUREMENT

Two elements are considered by OSHA in measuring the effectiveness of an organization's safety program: the frequency rate and the severity rate.

The frequency rate is the measurement of the number of disabling injuries per 200,000 worker-hours of exposure; the severity rate is the number of days lost per 200,000 worker-hours of exposure. Chapter 13, "Evaluating Safety Training Effectiveness," discusses a variety of other measures, as well as statistical techniques and some examples of each.

There are four important steps required by the OSHA recordkeeping system:

- Obtain a report on every injury requiring medical treatment.
- Record each injury on OSHA form 200 according to the instructions provided.
- Prepare a supplementary record of occupational injuries and illnesses on recordable cases, either on OSHA Form 101 or workmen's compensation reports giving the same information.
- Maintain these records on file for at least five years.

An OSHA recordable injury or illness is one that meets one or more of the following criteria:

- Ends in a fatality.
- Requires medical treatment other than first aid.
- Involves loss of consciousness.
- Involves a restriction of work or motion.
- Results in transfer to another job.

INTERNAL DATA ANALYSIS

During the year, periodically review the safety records to see where injuries are occurring and in what numbers. Look for any patterns or repeat situations. These records can help identify the high risk areas of a business that should receive immediate attention.

Even better, since the basic OSHA records include only injuries and illnesses, consider expanding the system to include all "incidents," including those where no injury or illness resulted, if such information would be of assistance in pinpointing unsafe conditions or procedures. Safety councils, insurance carriers, and others can also be of assistance in instituting such a system.

Injury/illness recordkeeping makes sense, and this practice is recommended to all employers. However as a result of recent amendments to the OSHAct of 1970, businesses are not required to keep records under the OSHA injury/illness recordkeeping system if they employ 10 or fewer employees. Employers with 11 or more employees at any one time during the year are required to maintain the injury/illness records.

EXTERNAL COMPARISONS

Regardless of the number of employees, a company may be selected by the federal Bureau of Labor Statistics (BLS) or a related state agency for inclusion in an annual sample survey. Selected companies will receive a letter directly from the agency with detailed instructions. Cooperation in the survey is expected.

In an effort to compare a company's accident frequency and severity rates to those of companies in comparable fields—that is, in a similar standard industrial code (SIC) classification—two sources of information are recommended:

- *Accident Facts,* most recent annual edition, published by the National Safety Council.
- *Chartbook on Occupational Injuries and Illnesses,* published by the Bureau of Labor Statistics.

7
How-To's in Specific Safety Training Techniques

SUITING UP FOR SAFETY

Background

In many cases the use of personal protective equipment is a safety procedure of last resort. It is worn when safety has not been designed into a machine or operation, or a hazard designed out of it. In other instances personal protective equipment is required to protect employees from themselves—for example, safety shoes in case someone drops a heavy wrench on his or her toe.

Regardless of whether personal protective equipment is a last resort procedure, however, the employer, as provided for under the OSHAct, must fully understand its intended purpose and how this relates to job procedures, and this information must be conveyed to the employees.

OSHAct's Subpart I, "Personal Protective Equipment," is not comprehensive. In most categories the reader is referred to the relevant ANSI standard.

The first section, 1910.132 ("General Requirements"), states that protective equipment—for eyes, face, head, and extremities; protective clothing; respiratory devices; shields and barriers—is considered to be personal protective equipment, and that such items shall be provided, used, and maintained where circumstances warrant.

It also says that where employees provide their own equipment, the employer is responsible for its adequacy, maintenance, and sanitation.

The section further states: "all personal protective equipment shall be of safe design and construction for the work to be performed."

In 1976 there were approximately 2.2 million disabling work injuries in the United States. Of these, about 12,500 were fatal and 80,000 resulted in some permanent disability. The approximate numbers of injuries involving different parts of the body, as estimated from the combined reports of state labor departments, are as follows:

Trunk	600,000
Fingers	350,000
Legs	280,000
Arms	200,000
General	180,000
Eyes, feet, hands, head (except eyes)	130,000
Toes	70,000

It can be assumed that at least some of these injuries could have been avoided if appropriate, clean, well-maintained protective equipment had been worn.

In an effort to increase the likelihood that such equipment will be worn, it is necessary to provide all employees with suitable and timely training regarding its adequacy, maintenance, and sanitation.

Additionally, all employees should be completely familiar with the following matters:

- Company policy on payment for personal protective equipment.
- Every area where any type of protective equipment must be worn.
- Company policy on violations of rules regarding personal protective equipment.
- Company policy on issuing, repair, and replacement of equipment.

Employee resistance to the use of personal protective equipment can be significantly reduced through the development of a positive attitude about its use. Motivational training on this subject is easy to present initially, but desired employee attitudes are difficult to maintain over the long haul.

Planning, Implementing, and Maintaining the Use of Protective Equipment

This section reviews ways to help plan, implement, and maintain personal protective equipment. This can be considered in terms of the following nine phases: (1) need analysis, (2) equipment selection, (3) program communications, (4) training, (5) fitting and adjustment, (6) target date setting, (7) break-in period, (8) enforcement, and (9) follow-through.

The first phase of promoting the use of personal protective equipment is called NEED ANALYSIS. Before selecting protective equipment, the hazards or conditions the equipment must protect the employee from must be determined. To accomplish this, questions such as the following must be asked:

- What standards does the law require for this type of work in this type of environment?
- What needs do our accident statistics point to?
- What hazards have we found in our safety and/or health inspections?
- What needs show up in our job analysis and job observations activities?
- Where is the potential for accidents, injuries, illnesses, and damage?
- Which hazards can't be eliminated or segregated?

The second phase of promoting the use of protective equipment is EQUIPMENT SELECTION. Once a need has been established, proper equipment must be selected. Basic consideration should include the following:

- Conformity to the standards.
- Degree of protection provided.
- Relative cost.
- Ease of use and maintenance.
- Relative comfort.

The third phase is PROGRAM COMMUNICATION. It is not appropriate to simply announce a protective equipment program, put it into effect, and expect to get immediate cooperation. Employees tend to resist change unless they see it as necessary, comfortable, or reasonable. It is helpful to use various approaches to publicity and promotion to teach employees why the equipment is necessary. Various points can be covered in supervisor's meetings, in safety meetings, by posters, on bulletin boards, in special meetings, and in casual conversation. Gradually, employees will come to expect or to request protective equipment to be used on the job. The main points in program communication are to educate employees in why protective equipment is necessary, and to motivate them to want it and use it.

TRAINING is an essential step in making sure protective equipment will be used properly. The employees should learn why the equipment is necessary, when it must be used, who must use it, where it is required, what the benefits are, and how to use it and take care of it. Don't forget that employee turnover will bring new employees into the work area. Therefore you will continually need to train new employees in the use of the protective equipment they will handle.

After the training phase comes the FITTING AND ADJUSTMENT phase. Unless the protective equipment fits the individual properly, it may not give the necessary protection.

There are many ways to fit or to adjust protective equipment. For example, face masks have straps that hold them snug against the contours of the head and face and prevent leaks; rubberized garments have snaps or ties that can be drawn up snugly, to keep loose and floppy garments from getting caught in machinery. The important point is to make sure the personal protective equipment really fits the employee, or else it will not provide the protection it is intended to give.

The next phase is TARGET DATE SETTING. After the other phases have been completed, set specific dates for completion of the various phases. For example, all employees shall be fitted with protective equipment before a certain date; all training shall be completed by a certain date; after a certain date, all employees must wear their protective equipment while in the production area.

After setting the target dates, expect a BREAK-IN PERIOD. There will usually be a period of psychological adjustment whenever a new personal protective program is established. Remember two things:

- You can expect some gripes, grumbles, and problems.
- Appropriate consideration must be given to each individual problem, then strive toward a workable solution.

It might also be wise to post signs that indicate the type of equipment needed. For example, a sign might read "EYE PROTECTION MUST BE WORN IN THIS AREA."

After the break-in phase comes ENFORCEMENT. If all the previous phases were successful, problems in terms of enforcement should be few. In case disciplinary action is required, sound judgment must be used and each case evaluated on an individual basis. Some employers follow the rules below for disciplinary action.

- First offense: an oral warning.
- Second offense: a written warning.
- Third offense: one week's suspension.
- Fourth offense: two week's suspension.
- Fifth offense: discharge.

These actions may seem harsh, but if employees fail to use protective equipment, they may be exposed to hazards. Don't forget, the employer can be penalized if employees do not use their protection.

The final phase is FOLLOW-THROUGH. Although disciplinary action may sometimes be necessary, positive motivation plays a more effective part in a successful protective equipment program. One type of positive motivation is a proper example set by management. Managers must wear their protective equipment, just as employees are expected to wear theirs.

Using protective equipment as a subject of safety talks and demonstrations can also have a positive motivating effect. Positive motivation can be gained from including personal protection in themes (carefully designed buttons, badges, banners, and brochures) and in contests that can lead to awards for those with exemplary performance in the use of protective equipment. Additional positive motivating forces are safety clubs that give recognition to employees for avoiding serious injury by proper use of their personal protection. A few examples of safety clubs that are active in certain establishments are as follows:

- Turtle Club: hard hats.
- Scarab Club: safety shoes.
- Halfway to Hell Club: safety nets.
- Wise Owl Club: eye protection.

Adequacy, Maintenance, and Sanitation

Before selling safety shoes and supplying safety goggles at a company store, the attendants must be guided by a well-structured program of equipment maintenance, preferably preventive maintenance.

Daily maintenance of different types of equipment might include adjustment of the suspension system on a safety hat; cleaning of goggle lenses, glasses, or spectacles; scraping residue from the sole of a safety shoe; proper adjustment of a face mask when donning an air-purifying respirator.

Performing these functions should be coupled with periodic inspections for weaknesses or defects in the equipment. How often this type of check is made, of course, depends on the particular type of equipment used. For example, sealed-canister gas masks should be weighed on receipt from the manufacturer, and the weight should be marked indelibly on each canister. Stored units should then be reweighed periodically, and those exceeding a recommended weight should be discarded even though the seal remains unbroken.

Sanitation, as spelled out in OSHAct, is a key part of any operation, and it requires the use of personal protective equipment, not only to eliminate cross-infection among users of the same unit of equipment, but because unsanitary equipment is objectionable to the wearer.

Procedures and facilities that are necessary to sanitize or disinfect equipment can be an integral part of an equipment maintenance program. For example, OSHAct says, "Respirators used routinely shall be inspected during cleaning." Without grime and dirt to hinder an inspection, gauges can be read better, rubber or elastomer parts can be checked for pliability and signs of deterioration, and valves can be checked.

Eye and Face Protection

When confronted with employee comments such as, "it's too tight," "it pinches my nose," or "these glasses blur my vision," safety professionals may throw up their hands in despair over the plant eye safety program.

To meet the requirements of the OSHAct regulation on eye and face protection, both employer and employee must cooperate: the employer must "make conveniently available a type of protector suitable for the work to be performed," and the employee "shall use such protectors."

To encourage this "use" by the employee, the law goes into detail about how the eyewear should fit and how it should be maintained. Provisions are made for employees who must have their vision corrected to perform their jobs. Phrases such as "reasonably comfortable," "fit snugly," and "easily cleanable" dot the wording of the regulation.

With this legislative weight, the safety professional has a strong selling point for the eyewear program. In addition to an explanation on equipment designed for specific types of hazard protection, a selection and fitting program is necessary. The elaborateness of such a program depends, of course, on the types of protection needed and the size of the work force. However informed and trained personnel should be used regardless of the size of the program. Not everyone knows how to use a pair of optical pliers to fit safety glasses, nor can just anyone spot allergic reactions to certain types of metal frame. But information such as "scratched or pitted lenses are injurious to vision, and can alter impact resistance" can be conveyed to all employees in the eyewear program by the safety professional. Self-monitoring preventive maintenance is much simpler to promote when employees are well informed of their responsibilities in a protective equipment program.

"Eye and Face Protection," CFR 1910.133, establishes the general requirements that no unprotected person shall knowingly be subjected to a hazardous condition and that "suitable eye protectors" shall be provided in the presence of flying objects, glare, liquids, injurious radiation, or any combinations of these.

Seven minimum requirements are set, as follows:

- Adequate protection against the particular hazards the device was designed for.
- Reasonably comfortable when worn under the designated conditions.
- Snug fit and no undue interference with the wearer's movements.
- Durability.
- Capable of being disinfected.
- Easily cleanable.
- Kept clean and in good repair.

The section also specifies that persons who need optical correction and eye protection shall wear either corrective eye protectors or goggles that can be worn over the protective lens. Paragraph 1910.133 also says that use will be in compliance with ANSI Z87.1-1978, "Occupational and Educational Eye and Face Protection." Other requirements concern the manufacturer.

Hand Protection

The OSHAct subpart on personal protective equipment has only one brief regulation referring to hand protection. Other sections of the law touch on hand safety with reference to tools and machine guarding, but Section 1910.137 requires rubber insulating gloves for electrical workers.

More standards are needed to cover the myriad industrial procedures in which hands require the protection of gloves. Items such as finger cots, head- and abrasion-resistant hand pads, and barrier creams, not to mention leather, cloth, asbestos, rubber, and neoprene gloves, are all forms of hand protection. Industrial operations meriting their use will eventually be spelled out. For example, nagging (and sometimes painful) itch of industrial dermatitis can be effectively counteracted with barrier creams and appropriate gloves.

Preemployment physicals aren't meant to analyze a person's health in depth, but a predisposition to allergic reactions and other diseases can be recorded. Based on such information, appropriate hand protection can be assigned to the worker who may react to any chemicals or solutions issued on the job. (Visual acuity, lifting restriction, and hearing problems are other areas checked during a preemployment physical that can also determine what other types of personal protecive equipment are assigned to a worker.)

There is no all-purpose form of hand protection. The material used for gloves depends largely on what is being handled.

- For most light work, a canvas or other cloth glove is both satisfactory and inexpensive.
- For rough or abrasive material, leather or leather-reinforced with metal stitching is required. Leather reinforced by metal stitching or metal mesh also provides good protection from edged tools.
- There are many plastic and plastic-coated gloves available, and they are designed to give protection from a variety of hazards. Some surpass leather in wearing ability. Others have granules or rough materials incorporated in the plastic for better gripping ability. Some are disposable.
- Asbestos gloves and combination gloves are available for handling hot materials.
- Metal mesh, terry cloth, and a variety of other glove constructions are available in a variety of styles to match hand protection with the particular hazard.
- Finger stalls of various materials, in combinations of one or more fingers, are available for situations that do not require a complete glove.
- Gloves coated with rubber, synthetic elastomers, polyvinyl chloride, or other synthetics offer protection against all types of petroleum products, caustic soda, tannic acid, and hydrochloric acid. They are also recommended in the handling of sulfuric acid. These gloves are available in varying degrees of strength to meet individual conditions.

Barrier creams, if properly used and reapplied frequently, provide limited protection against irritants to the hands and arms. There are four common types based on function—soap base, solvent-repellent, water-repellent, and special types. No one cream is effective against all irritants. The need for repeated washing to remove the barrier cream is one principal benefit from its use.

General Body Protection

OSHAct regulations stipulate that protective clothing shall be provided, used, and maintained. Yet no specific standards are listed for clothing that gives general body protection, such as jackets, aprons, trousers, lab coats, and lifelines.

With the advent of disposable clothing, a number of manufacturers now offer product lines suitable for use in clean-room operations. Some garments have antistatic and flame-retardant qualities that allow for repeated launderings. Other types of apparel are for one-time use only.

The design of garments such as leggings and sleeves that augment other types of protective equipment—gloves, respirators, boots—should facilitate wearing the basic protective devices or equipment. Sleeves should not be so tight fitting at the wrist that gloves cannot be worn comfortably and properly. Leggings should fit close to the leg so not to catch in moving machinery.

Lifelines, which are worn by employees who work above grade, should be inspected at regular intervals, and all general body protective clothing should be chosen so it can be put on and removed quickly.

Life vests should be worn by those who work over water.

Respiratory Protection

As a part of a human body increases in importance to sustaining life, the OSHAct regulations governing its protection increase in number. With a preliminary reference to the primary objective of "preventing atmospheric contamination," the law notes that such engineering controls are not always feasible, and there follows a series of explicit regulations on the protection of the respiratory system.

Again stipulating cooperation in the form of "employer shall provide" and "employee shall use," the law requires the establishment of written operating procedures as a part of a minimal acceptable program.

To assure adequate protection for the wearer of respiratory equipment, the person must be physically capable of performing the job functions while so protected. Periodic reviews are to be made of the medical status of the respirator user.

Once a worker has a clean bill of health to do a job while wearing a respirator, the OSHAct requires that he or she be trained in the use of the equipment and in its limitations. Here the safety professional must choose carefully the way to explain operation of the equipment. It is necessary to refrain from instilling fear of the hazard in the worker without conveying a false sense of security. By pointing out to the employee that the employer must supplement the equipment's use with surveillance and inspections of work area conditions, the safety professional can reinforce a positive attitude about the hazards of a job function that requires respiratory protection.

In addition to a number of regulations on the mechanical parts and air supplies for respirators, the law notes that "possible emergency and routine uses of respirators should be anticipated and planned for." The buddy system is required for areas where a wearer could be overcome by a toxic or oxygen-deficient atmosphere if the respirator failed. Communications (visual, voice, or signal line) are required between both or all individuals present.

In some plants where respiratory hazards are not part of daily operations, respirators are supplied for emergency use. The OSHAct requires that records be kept of equipment inspections, and stipulated maintenance procedures must be followed after each use.

"Respiratory Protection," CFR 1910.134, states that in control of occupational diseases caused by breathing air contaminated with harmful dusts, fogs, fumes, mists, gases, smokes, sprays, or vapors, the primary objective is to prevent atmospheric contamination. This is to be accomplished as far as feasible by accepted engineering control measures (e.g., enclosure or confinement of the operation, general and local ventilation, and substitution of less toxic materials). When effective engineering controls are not feasible, or while they are being instituted, appropriate respirators are to be used pursuant to the following requirements:

- Respirators shall be provided by the employer when such equipment is necessary to protect the health of the employee.
- The employer shall provide the respirators that are available and suitable for the purpose intended.
- The employer shall be responsible for the establishment and maintenance of a respiratory protective program.
- The employee shall use the provided respiratory protection in accordance with instructions and training received.

The requirements for a minimal program are as follows:

- Written standard operating procedures governing the selection and use of respirators shall be established.
- Respirators shall be selected on the basis of hazards to which the worker is exposed.
- The user shall be instructed and trained in the proper use of respirators and their limitations.

Suiting Up for Safety

- Where practicable, the respirators should be assigned to individual workers for their exclusive use.
- Respirators shall be regularly cleaned and disinfected. Those issued for the exclusive use of one worker should be cleaned after each day's use, or more often if necessary. Those used by more than one worker shall be thoroughly cleaned and disinfected after each use.
- Respirators shall be stored in a convenient, clean, and sanitary location.
- Respirators used routinely shall be inspected during cleaning. Worn or deteriorated parts shall be replaced. Respirators for emergency use, such as self-contained devices, shall be thoroughly inspected at least once a month and after each use.
- Appropriate surveillance of work area conditions and degree of employee exposure or stress shall be maintained.
- There shall be regular inspection and evaluation to determine the continued effectiveness of the program.
- Persons should not be assigned to tasks requiring use of respirators unless it has been determined that they are physically able to perform the work and use the equipment. The local physician shall determine what health and physical conditions are pertinent. The respirator user's medical status should be reviewed periodically (e.g., annually).
- Approved or accepted respirators shall be used when they are available. The respirator furnished shall provide adequate respiratory protection against the particular hazard for which it is designed in accordance with standards established by competent authorities.

Proper selection of respirators shall be made according to the guidance of American National Standard "Practices for Respiratory Protection" (Z88.2-1969).

Head Protection

The OSHAct requires protective helmets to meet the requirements of ANS Z89.1-1966 for impact and penetration resistance from falling and flying objects, but it is up to the safety professional and appropriate supervisors to make sure that the equipment is worn in the recommended manner.

To counter any deliberate employee mutilation (or misuse) of a safety helmet or bump cap, as frequently happens, the cost of replacement can be borne by the employee. The cutting or removal of suspension straps does not make a $15\frac{1}{2}$ ounce helmet (maximum weight as defined by ANS Z89.1) any more comfortable, and the effectiveness of the protection is diminished. Supervisors should be instructed by the safety professional to be thorough in their inspection for this type of equipment damage.

One type of head protection not provided for in the OSHAct is the protection of hair. flame-resistant net-type caps (where necessary), disposable nets, bandeaus, and other types of lightweight caps are available for both male and female industrial workers. These types of protection should be worn by any employee who choose to wear their hair in longer styles and whose hair is subject to entanglement in moving machinery or to ignition from sparks.

"Occupational Head Protection," CFR 1910.135, is very brief and merely refers the reader to ANS Z89.1-1969, "Safety Requirements for Industrial Head Protection."

The standard is intended primarily to guide manufacturers of hard hats in the construction and testing of protective headgear. But it might be well to know the characteristics of hats made to this standard as an aid in selection.

The standard divides hats into two types:

- **Type 1.** A helmet with a full brim at least 1½ inches wide.
- **Type 2.** A cap with no brim, but with a peak extending forward from the crown over the eyes only.
- Each type is divided into four classes:
  - *Class A.* Provides limited voltage protection.
  - *Class B.* Provides greater voltage protection.
  - *Class C.* Provides no voltage protection.
  - *Class D.* Provides limited voltage protection and is designed for fire fighting—it exists only in the form of Type 1.

All classes must provide the same protection from impact—that is, the helmet and its suspension may not transmit an average force of more than 850 foot-pounds, and no individual specimen may transmit more than 1000 foot-pounds when tested under prescribe standard conditions at temperatures of both 0 and 120°F.

Penetration allowable is slightly more for Class C helmets than for other classes.

Classes A and D, when tested under standard conditions at 2000 volts AC for one minute, must allow leakage of no more than 3 milliamperes. Class B (under ANS Z2.1, which is still in effect for this class) must take 15,000 volts AC for one minute with leakage of no more than 8 milliamperes. Class C helmets need offer no insulation from electric current.

Classes A and B must be of slow-burning material, and Class D must be of material that does not support combustion. Class C helmets are not tested for flammability, since they are made of metal.

All classes must restrict absorption of water under standard test conditions to no more than 5% of the weight of the helmet.

Standards call for the prominent labeling of all helmets with the letter indicating their class.

It is important to remember that the protection offered by a safety hat or cap depends on both the helmet shell and the suspension that separates it from the head. A defective or badly adjusted suspension permits the transmission of excessive force from the hard shell to the head.

Suspensions are not interchangeable between Class B helmets and other classes. Class B suspensions must include no metal parts and may not be supported by fittings that require the drilling of holes through the helmet.

Size is of critical importance in suspensions. Some companies use only nonadjustable suspensions, to assure against misadjustment by the worker. But whether the adjustable of nonadjustable type is chosen, the suspension in use must be right for the individual who wears it.

The protective quality of a hat or cap can be assured only by purchasing it from a reliable manufacturer or supplier and by careful inspection before it is put in use.

Finally, the best hat or cap in the world can deteriorate in use both from normal wear and tear and from lack of proper care and maintenance. Surveys have shown, for example, that many helmet shells are used after they have developed cracks, pits, or dents, which diminish their protective quality. Suspensions can become torn, frayed, or filthy.

Although OSHAct regulations make the employer responsible for proper hard hat maintenance, it does not define "proper maintenance." The following is general industry practice on hard hat maintenance.

- Good maintenance begins with the worker who uses the hat or cap. He or she should be provided with a secure, clean place to store headgear, where it is protected against blows, dirt, temperature extremes, and harmful chemicals. This is especially important when workers carry the headgear in trucks or cars. A safety cap or hat bouncing around with other equipment on the bed of a truck is not likely to remain in good condition long. A common place for carrying caps and hats in a car is the shelf behind the back seat. Here it is subject to damage and to excessive heat when the car is parked on a sunny day. Here, also it is likely to be turned into a missile if the car stops suddenly.
- Every part of each safety hat and cap should be visually inspected daily for signs of wear, damage, cracks, dents, pits, or other flaws.

Inspection should include the helmet shell, the suspension, the sweat band, and the accessories. Any headgear with defects that reduce its protective quality should be removed from service and replaced. Correction of the fault may be possible, but the headgear should not be used until correction is made.

- Cleanliness is of major importance, both to protect the worker against infection and irritation and to encourage the regular use of the protection by keeping it attractive. Companies should secure from the supplier or manufacturer instructions for cleaning both shell and suspension by means that are effective but do not damage the material. The responsibility for cleaning and sterilizing cannot be safely left to the individual employee.
- When hats or caps are shifted from one employee to another, the helmet should be carefully cleaned and the suspension sterilized or replaced with a new one.
- Sometimes it seems desirable to either the employer or the employee to paint the headgear. Check before allowing this, because some paints are incompatible with some plastic shell materials. It is usually better to purchase shells of colored plastic than to paint the desired color on the shell.
- Headgear can only protect if it is worn. This means that the employer must plan a program to gain employee acceptance of the need for safety headgear and to assure its regular and continued use.

The need for head protection should be carefully and clearly explained. The kinds of hazard that exist on the job and from which headgear offers protection should be presented clearly and forcefully. The effectiveness of safety hats and caps in preventing injuries should be made clear, but always with the honest acknowledgment that safety headgear, like any type of personal protective equipment, its a last line of defense when engineering and work procedures fail to prevent dangerous events.

It should be stressed that hats and caps protect against more than falling objects. Workers should understand that safety headgear may prevent injuries when the head strikes a fixed object, when it is struck by a moving but not falling object (crane hook, tool, etc.), when contact is made with electrical conductors, and when the worker is exposed to splashes of hot material or harmful chemicals.

Emphasis on these points will, in many situations, explain the need for a particular type of safety headgear. The need for a brimmed hat rather than a peaked cap, for example, may be made clear by stressing the likelihood of impacts other than those caused by objects falling vertically. The nature of the job will determine the degree of electrical protection needed, and this point should be thoroughly explained to the worker.

Hearing Protection

Although the OSHAct regulations set maximum permissible noise levels and exposures, and explain corrective procedures under Subpart G ("Occupational Health and Environment Control"), Section 1910.95, no mention is made in Subpart I ("Personal Protective Equipment") of hearing protection to be worn by the employee.

The Walsh-Healey Act and the references in the OSHAct are supplemented, for better understanding, by the Department of Labor's *Guidelines to Occupational Noise Exposure.*

Both Walsh-Healey and the OSHAct see engineering controls as the first line of defense against noise-induced hearing problems. However this method of resolving the problem of noise exposure is not always feasible. Hence the need for personal hearing protection equipment, although the government does not term this a permanent solution to the problem.

Beginning with the simple wax-impregnated cotton that the employee shapes and inserts in the ears, the selection of equipment continues through a diverse range, which includes muffs with features for coupling other types of personal protection with ear protection.

Selection of the most appropriate model of equipment depends on the type of noise problem in the employee's work area, the job operation to be performed, and to a certain degree, the individual's personal preference of equipment.

The preferred way to select the correct ear protection involves preemployment physicals (to determine any problems prior to on-the-job exposure) and the recommendation of an industrial hygienist who has surveyed the work area. Furthermore, only hearing protectors tested in accordance with the ANS Z24.22-1957, "Method for Measurement of the Real Ear Attenuation of Ear Protectors at Threshold," are acceptable for use according to the Department of Labor.

The seven-part "Directory of Noise Control Products," published in 1971 by National Safety News includes a section ("Personal Protection," National Safety News reprint stock no. 111.17-61) detailing models of ear protection as follows:

Inserts. Formed by user, premolded, and custom molded to block external ear canal from noise.

Superaural. Two disks covering ear canal openings and suspended from a headband.

Circumaural. Cup or dome-shaped devices placed over the ear, including the lobe, to reduce sound level before it reaches external auditory canal.

Whatever model of protection is selected, performance and comfort must be considered in addition to the application of the equipment to other job functions.

Educational and promotional programs on personal hearing protection are requirements for implementation of the Walsh-Healey Act and for an "effective hearing conservation program" as specified under the OSHAct, CFR 1910.95 paragraph (b)(3). It should be noted that standards are needed to establish minimum requirements for this type of personal protective equipment.

Regardless of the type of ear protector decided on, its attenuation, as stated by the manufacturer, must be sufficient to reduce the noise level in the worker's ear to the level and for the duration prescribed in Table 7-1. The manufacturer's stated values are determined under ideal conditions; therefore, as a precaution, it is wise to assume that the attenuation actually attained in use in the shop will be at least 5 decibels (dB) less than the stated value. Only ear protectors that have been tested in accordance with American National Standard Z24.22-1957, "Method for Measurement of the Real Ear Attenuation of Ear Protectors at Threshold," are acceptable to the Department of Labor.

The Department strongly recommends that any employee who is exposed to high sound levels and request ear protection be provided with it, even if the duration of exposure is within the limits prescribed by Table 7-1 and Figure 7-1.

When engineering and administrative controls fail to bring noise levels or duration of exposure to permissible levels, the use of personal protective equipment is required, as stated in the second sentence of paragraph b, Section 50-204.10 (Walsh-Healey Act) and CFR 1910.195:

If such controls fail to reduce sound levels within the levels of the table, personal protective equipment shall be provided and used to reduce sound levels within the levels of the table.

The use of personal protective equipment is considered by the Department to be an interim measure while engineering and administrative controls are being perfected. In very few cases will the use of this equipment be acceptable as a permanent solution to noise problems.

Some methods of control, such as providing an isolation booth for operators, or conducting noisy operations when few employees are in the

Table 7-1 Permissible Noise Exposure[a]

Duration per Day (hours)	Sound Level Slow Response (dB A)
8	90
6	92
4	95
3	97
2	100
1½	102
1	105
½	110
¼ or less	115

[a] Code of Federal Regulations, *Federal Register*, Washington, D.C. When the daily noise exposure is composed of two or more periods of noise exposure of different levels, their combined effect should be considered, rather than the individual effect of each. If the sum of the following fractions: $C_1/T_1 + C_2/T_2 + \ldots + Cn/Tn$ exceeds unity, then the mixed exposure should be considered to exceed the limit value. Cn indicates the total time of exposure at a specified noise level, and T_n indicates the total time of exposure permitted at that level.

plant, may require use of personal protective equipment by the operator when he or she must emerge from the booth to make adjustments, or by the few employees who carry on the noisy operation.

In addition, the regulations require that personal protective equipment be provided and used. How it is used is up to the employer. The Department recommends, however, that an educational and promotional program precede initiation of required use of such equipment, and continue as long as necessary to achieve 100% acceptance by employees. In the absence of an observable high proportion of use, the Department would consider the lack of a training and promotional program to constitute a violation of the regulation.

Foot Protection

Like Section 1910.135, Section 1910.136, "Occupational Foot Protection," does nothing but refer the reader to a standard—in this case ANS Z41.1-1967, "Standard for Men's Safety-Toe Footwear."

Plant Noise Survey Form

Date:
Meter #:

Location: _____

Number of personnel exposed: ____

Wearing ear protection? ☐ ☐
                                 Yes No

Operator: _____

Signed: _____

DIAGRAM:
(Show measuring location with an X.)

1565-A Procedure Checklist
1. Remove white cap.
2. Turn mode switch to "A_S" (or "C_F" for impact noise).
3. Turn attenuator knob to get scale reading (or to 130 for impact noise).
4. Add reading to knob setting to get total dB.

Notes: _____

	dB(A) max level	Exposure Time If necessary, note time of day	Total duration D (hours per day)	Permissible hours P per day	$\dfrac{D}{P}$
Maximum levels as measured on the "A_S" scale of the sound level meter. If the reading is between levels specified, use the higher level.	Over 115			None	
	115			$\frac{1}{4}$	
	110			$\frac{1}{2}$	
	105			1	
	102			$1\frac{1}{2}$	
	100			2	
	97			3	
	95			4	
	92			6	
	90			8	
	Under 90			Any	

Impact check (must be under 140 dB(C) on "C_F").

Total 0

To comply with regulations, total cannot be more than 1.

Recommendations: _____

Figure 7-1 Plant noise survey form.

Suiting Up for Safety

This standard outlines the specifications, performance requirements, and methods of tests for safety footwear designed primarily to protect the toe.

It divides shoes into three classes:

Class 75. This shoe must provide $16/32$-inch clearance after 75 foot-pounds of impact, and $16/32$-inch clearance after 2500 pounds compression.

Class 50. This shoe must provide $16/32$-inch clearance after 50 foot-pounds of impact and 1750 pounds compression.

Class 30. This shoe must provide $16/32$-inch clearance after 30 foot-pounds of impact and 1000 pounds compression.

The standard also provides that all shoes be clearly marked to indicate the class to which they belong.

Although not stated, it is implied that Class 75 is for heavy hazard exposure, Class 50 for medium exposure, and Class 30 for light exposure.

When selecting shoes, look for the identification of the class on the inside quarter, the shank of the outsole, shank of the insole, or tongue. The class number will be enclosed in a border.

The impact test for Class 75 calls for a 50-pound weight to be dropped from a height of 18 inches with at least $1/2$-inch clearance remaining between the toe cap and the insole of the shoe.

The compression test calls for applying a load, at the rate of 50 pounds per second, after the first 500 pounds of load, to the highest point of the toebox. The clearance between the toe cap and the insole must be at least $1/2$ inch.

Here again, OSHAct regulations make the employer responsible for maintenance but does not specify what constitutes proper maintenance.

In the case of steel-tipped shoes and other safety footwear, maintenance will be largely up to the individual.

Some steps management can take to ensure compliance are as follows:

Inform electricians with special shoes to avoid the use of nails in repairing.

Make periodic tests on conductive shoes to make sure the maximum allowable resistance (450,000 ohms) is not exceeded.

Carefully disinfect the boots after each shift when different shifts use the same boots.

Protection from Electrical Devices

Section 1910.137, "Electrical Protective Devices," refers the reader to existing standards, namely, ANS J6.1, J6.2, J6.4, J6.6, and J6.7, which cover rubber insulating gloves, matting, blankets, hoods, line hose, and sleeves. This section deals only with the items that are personal protective equipment—rubber insulating gloves and rubber insulating sleeves.

The ANSI standard covering rubber insulating gloves, J6.6, divides them into five classes.

> Class 0 with a minimum breakdown voltage of 6000 rms volts (AC) and 35,000 average volts (DC).
>
> Class 1 with a minimum breakdown voltage of 20,000 rms volts (AC) and 60,000 average volts (DC).
>
> Class 2 with a minimum breakdown voltage of 25,000 rms volts (AC) and 70,000 average volts (DC).
>
> Class 3 with a minimum breakdown voltage of 30,000 rms volts (AC) and 80,000 average volts (DC).
>
> Class 4 with a minimum breakdown voltage of 45,000 rms volts (AC) and 100,000 average volts (DC).

Maintenance of rubber gloves and sleeves consists mainly of washing. Mild soap flakes or a household detergent dissolved in warm water (about 90 to 100°F) can be used. Special compounds are also available commercially. Gloves and sleeves are then washed for about 15 minutes in a tumbler or agitator-type washer.

After washing the garments are rinsed in clear water and dried thoroughly. If a home-type dryer is used, the maximum temperature should be 150°F. A temperature of 120 to 130°F is recommended. But ozone-producing light bulbs should be removed from the dryer.

After inspection and retesting, the protective gear can be lightly dusted with soapstone or commercial talc and placed in stock.

All sleeves and gloves issued from storerooms or tool rooms should be visually checked for injury or damage that may have occurred in transit or storage. Particular attention should be given to equipment returning from special jobs (e.g., storm emergency).

Equipment showing tears, scratches, cuts, cracks, burns, or having a "dead feeling," should be tagged "defective" and taken out of service. The proper individual should then decide whether the item should be returned to laboratory for retest.

Suiting Up for Safety

Rubber gloves can be inspected by filling them with air or water and seeing if there are any leaks.

For rubber sleeves, stretching or rolling the rubber between the fingers will aid in revealing defects that normally can't be seen if the sleeve is lying flat.

As a general rule such equipment should not be stored when wet or dirty or in a manner than will distort or change a device from its natural shape. Rubber will take a permanent "set" if distorted from its original shape for a long time.

Protection from Environmental Hazards

Thus far OSHA has not promulgated any regulations dealing with control of two other types of industrial environmental hazards—temperature extremes and atmospheric pressure extremes. Responsible personnel should, however, be aware of the problems associated with these hazards.

Temperature Extremes

Extremely cold atmospheres are routine in meat processing, and they can cause problems in any industry that uses outdoor teams in cold climates.

The basic precaution against cold is proper clothing. It should be warm, allow perspiration, and not restrict circulation. Preferably there should be several layers of clothing. This provides layers of trapped air and air is a good insulator.

At below-zero temperatures workers should be instructed to wear gloves and avoid touching metal. Whenever possible, wood-handled tools should be provided.

If workers must work out-of-doors in very cold weather, they should be informed about the precautions and treatments associated with frostbite.

Whether working outdoors or in a deep-freeze, they should be cautioned to be careful while walking on frost and ice.

There are three approaches to contending with extreme heat—shielding, acclimatization, and personal protection.

Reflective screens, heat exchange screens, absorbing screens, and aluminized cloth curtains can be used to shield workers from radiant heat. In addition, a booth can be provided to shield personnel.

After about 10 days of relatively short daily exposures to a certain amount of heat, a person becomes acclimated. He or she is then able to perform a hot job with considerably less physical stress than an unacclimatized person would experience.

The beneficial effects of acclimatization are easily retained over a weekend, but they will be lost if there is no exposure to heat for a period of about two weeks.

Personal protection from extreme heat includes the vortex suit (which can also protect against cold) and reflective clothing.*

Atmospheric Pressure

Like temperature, atmospheric pressure must be dealt with at high and low extremes.

The main hazard of operations performed under high atmospheric pressure is decompression sickness or the "bends." Some of the controls of this condition include a properly equipped and manned decompression chamber, periodic medical checkups for workers, screening out of alcoholics, and use of an identification badge stating that the individual is a compressed air worker. (Because the symptoms of the bends resemble those of drunkenness, victims are often taken to jail rather than to a decompression chamber.)

Low atmospheric pressure is usually associated with high mountain elevations but is also found in some aerospace situations. The usual approach is pressurization or supplied-air masks.

Although these and other areas normally considered under the heading of industrial hygiene are not now covered by specific OSHAct regulations, subsequent directives may be issued that will spell out exacting procedures. Prudence requires employers to take advantage of whatever knowledge is available in these areas and to incorporate it in their industrial hygiene programs.

RESPONDING TO UNSAFE ACTS

Recognize Dangerous Activities

Accidents and injuries on the job don't just happen. They're the result of unsafe acts or conditions. Yet experience with hundreds of supervisors and thousands of employees in plants all over the United States proves that a competent supervisor can eliminate—in a reasonable and practical way—at least 50% of all unsafe acts and conditions. In other words, a supervisor is able to prevent half of all accidents and injuries on the job. How? By knowing how to recognize and check danger points.

---

*A vortex suit is a generic classification of a type of clothing that provides total body coverage (including head, hands, and feet) and that uses an external supply of circulating air that provides a constant ambient temperature for the wearer (e.g., astronaut's space suit).

"Unsafe acts" refers to employees' action. They fall into the category of human failure. On the other hand, "unsafe conditions" refer to the state of tools, machinery, clothing, or other inanimate objects. These are mechanical failures.

Failure to act safely will often lead to an unsafe condition. For example, an employee may forget to replace the cap on a drum of gasoline. This is an unsafe act. If the drum is allowed to remain uncovered, it sets up an unsafe condition. If fire breaks out as a result of this situation, it's a product of human and mechanical failure, having been caused by both an unsafe act and an unsafe condition.

Most accidents and injuries are caused by unsafe acts rather than by unsafe conditions. Roughly four accidents on the job are caused by unsafe acts for every one caused by unsafe conditions. Here's the breakdown of the causes of accident:

Mechanical failure (unsafe conditions)	20%
Human failure (unsafe acts)	78%
Acts of nature (floods, storms)	2%

What to Do When Unsafe Acts Occur

STOP the act immediately, consistently.

STUDY the job. If an employee commits an unsafe act because he or she thinks there is no other way to do the job, investigate the individual's work methods.

INSTRUCT. Once you decide how a job operation can be done more safely, instruct the employee in the correct method. Explain how the job should be done; demonstrate; then let the employee try it.

TRAIN an employee in the safe procedure. Check up on the person from time-to-time, making certain that he or she understands the safe procedure and will not go back to unsafe methods.

DISCIPLINE an employee only as a last resort, after the individual has repeatedly and willfully refused to follow safety rules.

Time Outs for Safety

How can an individual recognize an unsafe act when it occurs? Study the following checklist. It describes the possible categories of human failure that can result in an accident.

- **Operating Without Authority.** This includes any unauthorized action such as jumping on a moving vehicle, operating someone else's equipment without permission, and using tools or machinery for which the employee has not been trained.

- **Failure to Secure.** This refers to failure to tie down materials on a loaded vehicle, failure to lock or shut down switches, valves, or doors, and failure to shut off equipment when not in use.
- **Failure to Warn.** This includes failure of the employee to signal properly, failure to place warning signs or tags, and failure to take any action necessary to let others know that he or she is doing something that may put them in danger.
- **Operating at Unsafe Speed.** This includes actions such as running instead of walking, driving an automobile, truck, bus, or other vehicle above or below safe speeds, feeding or supplying production machines or assembly lines too slowly or too rapidly, throwing material instead of carrying or passing it, and using shortcuts that are unsafe.
- **Bypassing Safety Devices.** This includes disconnecting, removing, plugging, or blocking safety devices; failure to inspect signals, fuses, valves, and other safety devices, and to keep them in good repair; and ignoring signals, warning signs, tags, or other safety instructions.
- **Using Unsafe Equipment.** Using tools, machines, or materials that have become defective or otherwise made unsafe through wear and tear or abuse. This category also refers to the improper use of hands, feet, or other parts of the body in place of tools or machinery. It includes using safe equipment in an unsafe manner, such as gripping tools or other objects improperly or insecurely, or using the wrong equipment for a particular job.
- **Unsafe Loading.** Unsafe loading on a vehicle, platform, conveyor belt, or other apparatus means loading over the safe load limit, loading too high, or loading in such a way as to create a top-heavy load.
- **Unsafe Placing.** Unsafe placing refers to placing of tools, equipment, or other materials where they are in danger of rolling or falling, or where they become an obstruction in work areas, aisles, or other normal travel routes. It also refers to placing of hands in, on, or between pieces of equipment, or at dangerous points of operation.
- **Taking Unsafe Position or Posture.** Among such actions are lifting or carrying loads improperly; lifting with the body in a twisted or awkward position; walking or working on unguarded beams, girders, scaffolds; riding on tailboards, on running boards of trucks, or riding in precarious positions; passing on grades and curves; entering enclosures that are unsafe because of gases, temperature, or exposed power lines; failure to use proper methods of ascending or descending

when working in high places; standing in the line of travel of falling or moving objects; and taking a position that obstructs the free movement of others.
- **Working on Dangerous or Moving Equipment.** This category includes oiling, cleaning, or adjusting equipment while it is in motion; working on electrically charged equipment without cutting power; getting on or off vehicles while they are in motion; welding or repairing equipment containing flammable or explosive substances without first cleaning and venting; and unnecessary handling of materials while they are being processed on moving machines or conveyor belts.
- **Horseplay.** This is rough, boisterous play in a work environment; unapproved or unauthorized activity while on or off the job, but on company property.

Unsafe Practices Checklist

Supervisor practices that must be eliminated are as follows:

Ignorance of safety rules and procedures.
Inadequate understanding of certain tools and machines.
Inadequate preliminary safety instruction.
Inadequate follow-up on safety practices.
Inadequate supervision of students and/or employees.
Erratic safety instruction and evaluation.
Permissive attitude toward horseplay.
Inadequate and/or erratic enforcement of safety laws and regulations.
Unwillingness to set "safety example," or lack of concern.
Inadequate checking of machines before permitting operation.
Inadequate communications to administration about safety needs.
Inadequate maintenance of safety records.
Toleration of employee practices that should be eliminated.
Engaging in "horseplay" or other nonsense.
Ignoring safety instruction.
Failing to develop a positive attitude toward safety.
Being improperly dressed for assigned task.
Hurrying to finish a job or taking shortcuts where safety is concerned.
Forgetting or avoiding the use of safety guards.

RECOGNIZING UNSAFE CONDITIONS: OSHA'S PRIORITY HAZARDS

It is always helpful if one knows where trouble is most likely to occur. The following are OSHA's most frequently cited violations of job safety and health standards as detected by OSHA compliance officers during their on-site inspections of businesses. Specific references to pertinent OSHA standards as well as recommended corrective actions are also included.

Number 1

Violation: National Electrical Code: missing ground plugs, frayed or improperly spliced cords.

OSHA Section: 1910.309 Subpart S, Federal Register page 23782.

Corrective action and specific conditions to follow:

1. General: Section 1910.309 adopts as a national consensus standard the National Electrical Code NFPA 70-1971.
2. Scope covered: the provisions of this subpart S cover electrical installations and utilization equipment installed or used within or on public and private buildings, structures and other premises.
3. Exemptions:
    (a) Installations in ships, watercraft, railway rolling stock, aircraft, or automotive vehicles.
    (b) Underground mines.
    (c) Installation for exclusive use of railroads, communications utilities, or electrical utilities.
4. Applicable sections of National Electrical Code: Section 250-59 (a) (b) and (c)—portable and/or cord connected and plug connected equipment, grounding method. Section 400-5—flexible cords and cables, splices.
5. Interpretation: OSHA uses National Electrical Code as standard for inspection purposes; therefore all electrical installations should be checked for conformity to this standard.

Number 2

Violation: Mechanical power transmission apparatus; failure to guard belts, gears.

OSHA Section: 1910.219; *Federal Register* page 23729.

Recognizing Unsafe Conditions: OSHA's Priority Hazards

Corrective action and specific conditions to follow:

1. Belt, rope, and chain drives—horizontal belts and ropes: where both runs of horizontal belts are 7 feet or less from the floor level, the guard shall extend at least 15 inches above the belt, except that where both runs of a horizontal belt are 42 inches or less from the floor, the belt shall be fully enclosed.
2. Overhead horizontal belts: overhead horizontal belts with lower parts 7 feet or less from the floor or platform, shall be guarded on sides and bottom.
3. Vertical and inclined belts shall be enclosed by a guard.
4. Standard guards shall be made from the following materials; expanded metal, perforated or solid sheet metal, wire mesh on a frame of angle iron, or iron pipe securely fastened to floor or to frame of machine.
5. Gears shall be completely enclosed or by a standard guard (described in 4), or by a hand guard covering the face of gear.
6. Interpretation: all belts and gears must be fully enclosed to prevent injury to personnel operating machinery.

Number 3

Violation: Portable fire extinguishers: improperly mounted or inaccessible extinguishers.

OSHA Section: 1910.157; *Federal Register* page 23683.

Corrective action and specific conditions to follow:

1. Extinguishers shall be conspicuously located where they will be readily accessible and immediately available in the event of fire. They shall be located along normal paths of travel.
2. Extinguishers shall not be obstructed or obscured from view. Where visibility may be obscured means shall be provided to indicate the location and intended use of extinguishers.
3. Extinguishers shall not weight in excess of 40 pounds and shall be mounted so that top of extinguisher is not more than 5 feet above the floor.
4. Extinguishers mounted in cabinets shall be placed in a manner such that the operating instructions face outward.
5. Extinguishers installed where subject to severe vibrations shall be installed in brackets specifically designed for this use.
6. Interpretation: fire extinguishers must be located in conspicious locations and should be mounted securely.

Number 4

Violation: General requirements for all machines: missing guards to protect against nip points, rotating parts, flying chips or sparks.

OSHA Section: 1910.212; *Federal Register* page 23712.

Corrective action and specific conditions to follow:
1. One or more methods of machine guarding shall be provided to protect the operator and other employees in the machine area from hazards such as those created by point of operation, ingoing nip points, rotating parts, flying chips and sparks.
2. Guards shall be attached to machine or secured elsewhere if necessary.
3. Guards shall not offer an accident hazard in itself.
4. Point of operation guards shall be constructed as to prevent the operator from having any part of his or her body in the danger zone during the operating cycle.
5. Interpretation: machinery shall be guarded to such an extent as to prevent injury to personnel from flying material or operation of machinery.

Number 5

Violation: Woodworking machinery: missing guards on saws.

OSHA Section: 1910.213; *Federal Register* page 23713.

Corrective action and specific conditions to follow:
1. All belts, pulleys, gears, shafts, and moving parts shall be guarded in accordance with corrective actions stated in violation number 2.
2. Portions of saw blade which a person may come into contact with shall be covered with an exhaust hood or a guard that shall be arranged as to prevent accidental contact with the saw.
3. Interpretation: woodcutting machinery must be shielded in such a way as to prevent injury to personnel from gears, pulleys, or the saw blade itself.

Number 6

Violation: General requirements—walking and working surfaces: failure to keep work areas clean, orderly, and sanitary.

OSHA Section: 1910.22; *Federal Register* page 23508.

Corrective action and specific conditions to follow:

1. All places of employment, passageways, storerooms, and service rooms shall be kept clean and orderly and in a sanitary condition.
2. The floor of every workroom shall be kept clean and if possible dry.
3. Where wet processes are used, drainage shall be maintained and false floors or other dry standing places shall be provided.
4. All floors and passageways shall be kept free from protruding nails, splinters, holes, or loose boards.
5. Interpretation: sanitary work conditions must be maintained for workers.

Number 7

Violation: Abrasive wheel machinery: missing guards, or tool tests not adjusted to within $1/8$ inch of wheel face.

OSHA Section: 1910.215; *Federal Register* page 23717.

Corrective action and specific conditions to follow:

1. Abrasive wheel machinery shall be provided which shall cover the spindle end, nut, and flange projections.
2. The safety guard shall be mounted to maintain proper alignment with the wheel and the strength of the fastenings shall exceed the strength of the guard.
3. Work rests shall be used to support the work. They shall be of rigid construction and adjustable to compensate for wheel wear.
4. Work rests shall be kept adjusted to within $1/8$ inch to prevent the work from jamming in machinery.
5. Interpretation: abrasive wheel machinery must have adequate guards to prevent injuries and work rest must be adjusted to $1/8$ inch maximum distance from abrasive wheel.

Number 8

Violation: Welding, cutting, and brazing: improper storage and handling of compressed gases.

OSHA Section: 1910.252; *Federal Register* page 23738.

Corrective action and specific conditions to follow:

1. Gas cylinders shall be kept away from radiators and other sources of heat.
2. Inside buildings, gas cylinders shall be stored in a well-protected, well-ventilated, dry location at least 20 feet from highly combustible materials.

3. Cylinders shall be stored in definitely assigned places away from elevators, stairs, and gangways.
4. Assigned storage areas shall be located where cylinders cannot be damaged by passing or falling objects.
5. Cylinders should be located away from areas subject to tampering by unauthorized persons.
6. Cylinders shall not be kept in unventilated enclosures.
7. Signs shall be conspicuously posted in such rooms reading: "Danger—No Smoking, Matches, or Open Lights."

Number 9

Violation: Powered industrial trucks: lift trucks left running and unattended.

OSHA Section: 1910.178; *Federal Register* page 23696.

Corrective action and specific conditions to follow:
1. When a powered industrial truck is left unattended, load engaging means shall be fully lowered, controls shall be neutralized, power shall be shut off, and brake set. Wheels shall be blocked if the truck is parked on an incline.
2. A powered industrial truck is unattended if the operator is 25 feet or more away from the vehicle which remains in his or her view or whenever the operator leaves the vehicle and it is not in view.
3. Interpretation: a powered industrial truck is unattended if the operator is 25 feet or more away from his or her vehicle or is out of sight of the vehicle. When a lift truck is left unattended it must be incapable of movement.

Number 10

Violation: Means of egress, general: unmarked exits.

OSHA Section: 1910.37 Subpart G; *Federal Register* page 23679.

Corrective action and specific conditions to follow:
1. Exits shall be marked by a readily visible sign.
2. Doors, passages, or stairways which may be mistaken for an exit shall be marked "not an exit," or shall be identified as to purpose.
3. Signs must clearly be identified as stating EXIT, and not be masked or confused by surrounding environment.
4. Where an exit is not immediately apparent, a sign with an appropriate arrow indicating the direction to the exit shall be placed.

5. Illumination of exit sign shall not be less than 5 foot candles, 25 square inch minimum translucent area for internally illuminated signs.
6. An internally illuminated sign shall be placed in areas where normal background illumination may be reduced.
7. The word EXIT shall be identified in legible letters not less than 6 inches high, 3/4 inch minimum stroke.

Number 11

Violation: Flammable and combustible liquids: storing too much or uncovered containers of flammable or combustible material.

OSHA Section: 1910.106; *Federal Register* page 23616.

Corrective action and specific conditions to follow:
1. Container tanks used for storage of flammable or combustible liquids shall not exceed 60 gallons individual capacity.
2. Portable storage tanks used for the storage of flammable or combustible liquids shall not exceed 660 gallons individual capacity.
3. Flammable or combustible liquids shall be kept in covered containers when not actually in use.

Number 12

Violation: Sanitation: failure to provide a clean, sanitary wash place.

OSHA Section: 1910.141 Subpart J; *Federal Register* page 23673.

Corrective action and specific conditions to follow:
1. Potable water shall be provided in all places of employment, for drinking, washing of the person, cooking, washing of foods, washing of cooking or eating utensils, and personal service rooms.
2. Washing facilities shall be maintained in a sanitary condition.
3. Each lavatory shall be provided with hot and cold running water.
4. Hand soap or similar cleansing agents shall be provided.
5. Individual hand towels of cloth or paper, warm air blowers or clean individual sections of continuous cloth toweling, convenient to the lavatories, shall be provided.

Number 13

Violation: Spray finishing using flammable or combustible liquids: failure to post or observe "No Smoking" signs in paint spray areas.

OSHA Section: 1910.107; *Federal Register* page 23628.

Corrective action and specific conditions to follow:

1. "No Smoking" signs in large letters or contrasting color background shall be conspicuously posted at all spraying areas and paint storage rooms.

Number 14

Violation: Hand and portable power tools and equipment, general: missing guards.

OSHA Section: 1910.243; *Federal Register* page 23734.

Corrective action and specific conditions to follow:

1. All portable, power-driven circular saws having a blade diameter greater than 2 inches shall be equipped with guards above and below the base plate or shoe.
2. Portable belt sanding machines shall be provided with guards at each nip point where the sanding belt runs on to a pulley.
3. Interpretation: portable power tools with the exception of drills must be guarded to prevent personal injury.

Number 15

Violation: General requirements, means of egress: blocked or inadequate exits.

OSHA Section: 1910.37; *Federal Register* page 23531.

Corrective action and specific conditions to follow:

1. Every building shall be provided with exits of kinds, numbers, location, and capacity appropriate to the individual building or structure.
2. Exits shall be arranged and maintained as to provide free and unobstructed egress from all parts of the building.
3. No lock or fastening, to prevent free escape from the inside of any building shall be installed.
4. Every required exit way shall be continuously maintained free of all obstructions or impediments to full instant use.

Number 16

Violation: Portable wood ladders: missing rungs, weak or wobbly legs, missing braces.

OSHA Section: 1910.25; *Federal Register* page 23515.

Recognizing Unsafe Conditions: OSHA's Priority Hazards   111

Corrective action and specific conditions to follow:
1. Ladders shall be maintained in good condition at all times; the joint between the steps and side rails shall be tight, all hardware and fittings securely fastened.
2. The distance between rungs shall not exceed 12 inches and shall be uniform throughout the length of the ladder.

Number 17

Violation: Noise exposure: noise in excess of 90 decibels on an eight-hour time-weighted average.

OSHA Section: 1910.95; *Federal Register* page 23597.

Corrective action and specific conditions to follow:
1. Permissible noise exposure shall not exceed 90 decibels on an eight-hour time-weighted average.
2. When sound exceeds the permissible limits engineering controls shall be utilized to reduce sound levels.
3. If controls fail or controls have yet to be completed, personal protective equipment shall be provided and used to reduce sound levels to within permissible levels.

Number 18

Violation: Medical services and first aid: missing or inadequate first aid kit.

OSHA Section: 1910.151; *Federal Register* page 23682.

Corrective action and specific conditions to follow:
1. In the absence of an infirmary, clinic, or hospital in proximity to the workplace which is used for the treatment of all injured employees, a person or persons shall be adequately trained to render first aid.
2. First aid supplies approved by a consulting physician shall be readily available.

Number 19

Violation: Eye and face protection: failure to wear protective goggles where there is strong possibility of danger to the eyes from flying objects.

OSHA Section: 1910.133; *Federal Register* page 23670.

Corrective action and specific conditions to follow:

1. Protective eye and face equipment shall be required where there is a reasonable probability of injury that can be prevented by such equipment.
2. Protectors shall provide adequate protection against the particular hazards for which they were designed.
3. They shall fit snugly and shall not unduly interfere with the movements of the wearer.
4. Design, construction, testing, and use of devices for eye and face protection shall be in accordance with American National Standard for Occupational and Educational Eye and Face Protection.

Number 20

Violation: Personal protective equipment, general: missing hard hats where there is danger of dropped items, striking against objects, or electrical shock.

OSHA Section: 1910.135; *Federal Register* page 23673.

Corrective action and specific conditions to follow:

1. Helmets for the protection of heads of occupational workers from impact and penetration from falling and flying objects and limited electrical shock shall be provided.
2. Helmet construction shall meet requirements and specifications established in American National Standard Safety Requirements for Industrial Head Protection.
3. Interpretation: helmets must be worn in areas where personal injury could be sustained from falling or flying objects. These helmets must be approved by American National Standard Safety Requirements for Industrial Head Protection.

Number 21

Violation: Fixed ladders: rungs too widely spaced; too little clearance to sides or rear of ladder.

OSHA Section: 1910.27; *Federal Register* page 23519.

Corrective action and specific conditions to follow:

1. The distance between rungs, cleats, and steps shall not exceed 12 inches and shall be uniform throughout the length of the ladder.
2. The perpendicular distance from the centerline of the rungs to the nearest permanent object on the climbing side of the ladder shall be 36 inches for a pitch of 76 degrees, and 30 inches for a pitch of 90 degrees.

3. The distance from the centerline of rungs to the nearest permanent object in back of the ladder shall be not less than 7 inches.

ELIMINATING UNSAFE CONDITIONS

The following steps should be taken to effectively and efficiently eliminate an unsafe condition:

- **Remove.** If at all possible, have the hazard eliminated.
- **Guard.** If danger point (i.e., high tension wires) can't be removed, see to it that hazard is shielded by screens, enclosures, or other guarding devices.
- **Warn.** If guarding is impossible or impractical, warn of the unsafe condition. If a truck must back up across a sidewalk to a loading platform, the sidewalk cannot be removed or a fence built around the truck. All that can be done is to warn that an unsafe condition exists. This is done by posting a danger sign or making use of a bell, horn, whistle, signal light, painted striped lines, red flag, or other device.
- **Recommend.** If you cannot remove or guard an unsafe condition on your own, notify the proper authorities about it. Make specific recommendations as to how the unsafe condition can be eliminated.
- **Follow up.** After a reasonable length of time, check to see whether the recommendation has been acted on, or whether the unsafe condition still exists. If it remains, the person or persons to whom the recommendations were made should be notified.

The following factors should be considered in organizing a plant that provides for maximum productivity and employee well-being:

- The general arrangement of the facility should be efficient, orderly, and neat.
- Work stations should be clearly identified so employees can be assigned according to the most effective working arrangement.
- Material flow should be designed to prevent unnecessary employee movement for given work.
- Materials storage, distribution, and handling should be routinized for efficiency and safety.
- Decentralized tool storage should be used wherever possible. Where centralized storage is essential (e.g., general supply areas, locker areas,

and project storage areas), care should be given to establish a management system that will avoid unnecessary crowding or congested traffic flow. (Certain procedures, such as time staggering may reduce congestion.)
- Time-use plans should be established for frequently used facilities to avoid having workers wait for a particular apparatus.
- A warning system and communications network should be established for emergencies such as fire, explosion, storm, injuries, and other events that would affect the well-being of employees.

The following unsafe conditions checklist presents a variety of undesirable charcteristics to which both employers and employees should be alert.

- **Unsafe Conditions—Mechanical Failure.** Here are types of unsafe conditions that can lead to occupational accidents and injuries. *Note.* Keep in mind that unsafe conditions often come about as a result of unsafe acts.
- **Lack of Guards.** This applies to hazardous places like platforms, catwalks, or scaffolds where no guard rails are provided; power lines or explosive materials that are not fenced off or enclosed in some way; and machines or other equipment where moving parts or other danger points are not safeguarded.
- **Inadequate Guards.** Often a hazard that is partially guarded is more dangerous than it would be if there were no guards. The employee, seeing some sort of guard, may feel secure and fail to take precautions that would ordinarily be taken if there were no guards at all.
- **Defects.** Equipment or materials that are worn, torn, cracked, broken, rusty, bent, sharp, or splintered; buildings, machines, or tools that have been condemned or are in disrepair.
- **Hazardous Arrangement (housekeeping).** Cluttered floors and work areas; improper layout of machines and other production facilities; blocked aisle space or fire exits; unsafely stored or piled tools and material; overloaded platforms and vehicles; inadequate drainage and disposal facilities for waste products.
- **Improper Illumination.** Insufficient light; too much light; lights of the wrong color; glare; arrangement of lighting systems that result in shadows and too much contrast.
- **Unsafe Ventilation.** Concentration of vapors, dusts, gases, fumes; unsuitable capacity, location, or arrangement of ventilation system;

insufficient air changes, impure air source used for air changes: abnormal temperatures and humidity.

Examples of Conditions To Be Inspected

In describing conditions for each item to be inspected, terms such as the following should be used:

Broken.	Leaking.
Corroded.	Loose (or slipping).
Decomposed.	Missing.
Frayed.	Rusted.
Fuming.	Spillage.
Gaseous.	Vibrating.
Jagged.	

Hazard Classification

It is important to differentiate the *degrees of severity* of different hazards. The commonly used standards are given below.

- **Class A Hazard.** Any condition or practice with *potential* for causing *loss* of life or body part and/or extensive loss of structure, equipment, or material.
- **Class B Hazard.** Any condition or practice with *potential* for causing serious injury, illness, or property damage, but less severe than Class A.
- **Class C Hazard.** Any condition or practice with *probable potential* for causing *nondisabling* injury or illness, or *nondisruptive* property damage.

ACCIDENT REPORT PREPARATION

Model Injury Reports

Included in a model injury report are the name and job title of the injured party, the time and date of the accident, information on where and how the accident occurred, and an assessment of the nature and extent of the injury. See Tables 7-2 and 7-3.

Table 7-2 Model Injury Reports: Accident Report 1[a]

WHO
    NAME OF INJURED ....
    JOB TITLE..
WHEN
    DATE OF ACCIDENT ....
    TIME OF ACCIDENT...
WHERE
    PLACE OF ACCIDENT ....
    ....
HOW
    CAUSE OF ACCIDENT ....
    ....
WHAT WAS NATURE AND EXTENT OF INJURY.......................

[a] Reprinted with permission from the Bureau of Business Practice, Waterford, CT 06386. Copyright, Bureau of Business Practice, Inc.

Table 7-3 Model Injury Reports: Accident Report 2[a]

Date of report............Made by ....Clock no. ....
Name of injured employee..
Employee's clock no......................Dept..............................
Occupation..............Date of injury............Time ....
Where did injury occur? ....
Describe injury in detail...
What caused injury?...
....
Is further treatment required? ....
Is hospital treatment required?...
Will injury cause loss of time? ....For how long? ....
When is employee expected to return to work? ....
Name of person giving initial treatment ....

[a] Reprinted with permission from the Bureau of Business Practice, Waterford, CT 06386. Copyright, Bureau of Business Practice, Inc.

Hot Surfaces—9%. Guard against exposure of combustibles to furnaces, hot ducts or flues, electric lamps or heating elements, and hot metal inprocess.

Overheated Materials—7%. Watch out for abnormal process temperatures, overheating of materials in driers, and overheating of flammable liquids.

Open Flames—6%. Be especially careful with gasoline or other torches. Stay alert for trouble with gas or oil burners.

Foreign Substances—5%. Caution operators to watch for foreign material in stock. Tramp metal produces sparks when struck by rapidly revolving tools. Oil pans used near cutting machines are a constant hazard.

Spontaneous Heating—4%. Watch for deposits in ducts and flues, low-grade storage, scrap waste, oily waste, and rubbish.

Cutting and Welding—4%. Supervise with care; these operations are highly dangerous in areas where sparks can ignite combustibles.

Combustion Sparks—4%. Instruct workers to be careful when burning rubbish; watch foundry cupolas, furnaces, and fireboxes.

Miscellaneous—10%. This includes incendiary cases (3%); fires spreading from adjoining buildings (2^1/$_2$%); molten metal or glass (2%); static electricity near flammable liquids, as at spreading or coating rolls or where liquid flows from pipes (1^1/$_2$%); chemical action (^{1}/$_2$%); and lightning (^{1}/$_2$%).

FIRE PREVENTION PROGRAM COMPONENTS

The following factors should be considered in implementing a fire prevention program:

Gas. Gases of all types (oxygen, acetylene, methane, propane, butane, etc.) represent a hazard. All equipment (piping, regulators, and ignitors) should receive periodic and careful inspection by a qualified person. Employees must receive thorough instruction before using these types of equipment. Matches should not be used for lighting gas-fired equipment.

Housekeeping. Rubbish, waste, and other debris must be cleaned up and removed daily and disposed of in suitable containers outside of the plant.

Electrical Equipment. Inspection and maintenance of motors and other electrical devices should be done by a qualified person. All electrical equipment should be included in a periodic inspection.

Matches and Smoking. Smoking and the use of matches should be permitted in designated areas only. Oily or paint-soaked rags or clothing left in lockers are a great hazard. Dust, lint, and oil collecting in ducts and flues can also ignite spontaneously. Safety containers should be provided for rags and waste, and they should be emptied daily. Employees must be instructed carefully and supervised closely to prevent fires of these types. Ashtrays for personal smoking materials (e.g., matches, cigarettes, and so on) should be conspicuously provided for employee use. Appropriate instruction in the use of these receptacles should reduce the likelihood of an accidental fire.

Open Flames. Open flames are involved in welding, forging, forming, and other heating operations. Fireproof materials around the work area and suitable clothing will aid in the prevention of most fires. However the use of portable equipment involves high fire danger and all precautions should be taken, including having an extinguisher of the proper type immediately available.

Heated Surfaces. Heated surfaces on furnaces, flues, heating devices, and light bulbs can be the cause of fires if flammable or combustible materials are close enough to absorb sufficient heat to cause combustion. Care should be taken to ensure that all such devices are properly installed, especially with respect to clearance and barrier materials.

Molten Metal. Molten metal will ignite any flammable material with which it comes in contact. Precautions should be taken to provide a flameproof environment in foundry areas and also to provide flameproof clothing.

Volatile Liquids. Paints, varnishes, lacquers, petroleum products, solvents, and similar volatile materials are frequent sources of fire and explosion. The liquids themselves do not burn, but heat causes vaporization a the vapors of these liquids are easily ignited by sparks from electrical de 's, static electricity, or metal objects striking together or, of course, op mes. Vapors are generally heavier than air and may travel along the for considerable distances before they are ignited and flash back to the ce. Tight metal containers, flameproof cabinets, color coding, and st orage and handling procedures are necessary. It is necessary that al ns in the plant know exactly what to do in the case of fire.

okers

te all c nd warnings to the contrary, some persons persist in
ig, evene lly designated areas in which smoking materials pre-
lear hs.

It is sometimes necessary to post signs indicating that smoking is or is not permitted. These signs should be discussed in training sessions and all employees made aware of the importance of self-imposed smoking discipline.

Smokers should be cognizant of various location and usuage restrictions that apply throughout the company's facilities.

Smokers who persist in violating company rules relative to smoking must be dismissed or transferred in an effort to protect the individuals, their fellow employees, and company property.

Emergency Response in the Event of Fire

Every employee must be given instruction on the following procedures:

How and when to turn in a fire alarm.

How to extinguish fires and how to determine whether extinguishment should be attempted.

What to do in the event of alarm (turn off all machines, disconnect power at main switch, turn off gas valves and torches, etc.).

How to leave the building.

A major fire is a disaster in any industry. Constant vigilance and quick action by the line supervisors can frequently contain a small blaze until the plant fire brigade or community fire department can extinguish it.

All supervisors are responsible for keeping areas around the fire doors and emergency exits clear at all times. Many plants have fire doors that close automatically to seal off an area to prevent the spreading of a fire. If these doors are blocked, however, they cannot function.

Another way supervisors can help to prevent a major fire is to make certain all employees know the location of all extinguishers and how to use them. This should be part of every indoctrination program for all new employees as well as the subject of regular review. Also be sure that everyone knows what type of extinguisher to use for a particular type of fire. Here's a rundown of fires and extinguishers of various types.

FIRES AND EXTINGUISHERS

Traditionally, fires have been classified by their sources or fuels. Depending on whether the source that fuels the fire is an ordinary combustible, a flammable liquid, electrical equipment, or certain elemental metals, a particular portable fire extinguisher is recommended by the National Fire Protection

Association (NFPA). Figure 8-1 presents the NFPA-approved classification system currently in use.

There are four major types of portable extinguishers available for purchase and use. Figure 8-2 summarizes these units by pressure source and method of operation.

Depending on the intended application of a fire extinguisher, it is necessary to be aware of the scope of products available for use. Figure 8-3 offers a guide to portable fire extinguishers.

No matter how extensive the testing of any mechanical device, there are still some hazards associated with their use, and operators of the devices should be advised of these hazards. Any training program related to the selection and usage of fire extinguishers should include material that describes their inherent dangers. See Figure 8-4.

Operation of a portable fire extinguisher is a relatively simple task if one has been properly and completely trained. The correct method to be used varies depending on the class of fire (A, B, C, or D) and the type of fire extinguisher (water, dry chemical-stored pressure, dry chemical-cartridge, or carbon dioxide). Figure 8-5 summarizes the selection and application of portable fire extinguishers by extending Figure 8-1.

Some recent attempts have been made to make fire extinguisher opera-

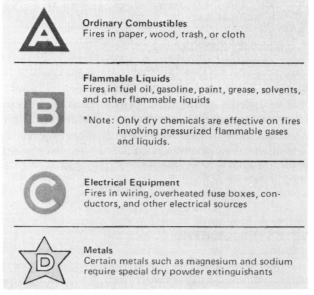

Figure 8-1 Comparative characteristics of fire extinguishers. Reprinted with permission from *Plant Engineering Magazine,* Technical Publishing, A Company of the Dun & Bradstreet Corporation—all rights reserved.

Fires and Extinguishers

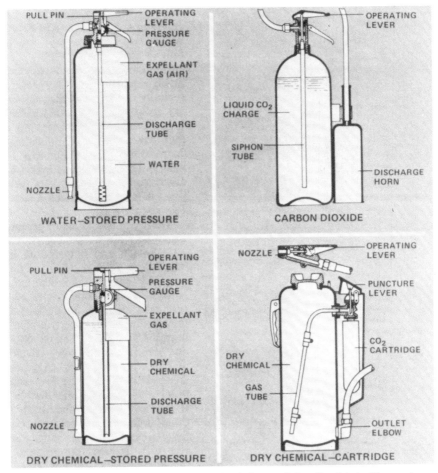

Figure 8-2 Portable fire extinguisher operational characteristics. Reprinted with permission from *Plant Engineering Magazine,* Technical Publishing, A Company of the Dun & Bradstreet Corporation—all rights reserved.

tion easier to understand. Figure 8-6 presents an example of newer instructions being tried to simplify fire extinguisher use.

Extinguisher Maintenance and Instruction

Take a careful inventory of the extinguishers in each area and be sure that they are maintained properly. Many supervisors supplement the company fire brigade by training their own workers and drilling them to keep them sharp and aware of their assignments. By doing this, and by assigning men

GUIDE TO PORTABLE FIRE EXTINGUISHERS

Class of fire	A		A/B	B/C				A/B/C		D
	Water types		AFFF Foam	Carbon dioxide	Dry chemical types		Halon 1211	Multipurpose dry chemical		Halon 1211	Dry powder
Type of unit	STORED PRESSURE	PUMP TANK	STORED PRESSURE	SELF EXPELLING	STORED PRESSURE	CARTRIDGE OPERATED	STORED PRESSURE	STORED PRESSURE	CARTRIDGE OPERATED	STORED PRESSURE	CARTRIDGE OPERATED
Sizes available	2½ Gal.	2½ and 5-Gal.	2½ Gal.	5-20 lb.	2½-30 lb.	4-30 lb.	2 to 22 lb.	2½-30 lb.	5-30 lb.	9 to 22 lb.	30 lb.
Approximate horizontal range	30 to 40 ft.	30 to 40 ft.	20 to 25 ft.	3-8 ft.	10-15 ft.	10-20 ft.	10 to 16 ft.	10-15 ft.	10-20 ft.	14 to 16 ft.	5 ft.
Approximate discharge time	1 Min.	1 to 3 Min.	50 Sec.	8-15 Sec.	8-25 Sec.	8-25 Sec.	8 to 18 Sec.	8-25 Sec.	8-25 Sec.	10 to 18 Sec.	20 Sec.

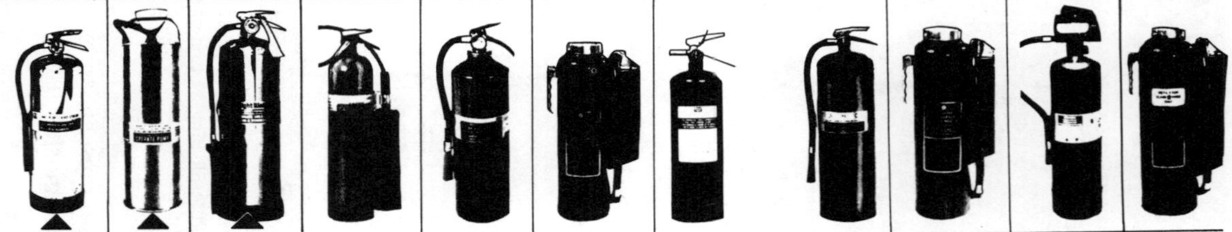

▲ Protection required below 40 F.

Figure 8-3 Guide to portable fire extinguishers. Reprinted with permission from *Plant Engineering Magazine*, Technical Publishing, A Company of the Dun & Bradstreet Corporation—all rights reserved.

THE DANGERS OF PORTABLE FIRE EXTINGUISHERS

Portable fire extinguishers essentially are safe to use. Manufacturers take steps during production to assure that a quality product is made; units and their components are tested before assembly. Most portable units are approved by Factory Mutual (FM) for the classes of fire for which they are suitable, or are listed by Underwriters Laboratories (UL), rated for class of fire and extinguishing potential. Many units bear the seals of both testing agencies, whose quality control representatives inspect manufacturers periodically, exercising continual, careful control over the products they have approved or listed.

Despite testing, approving, research, and quality control, some fire extinguishers pose potential dangers when they are damaged or improperly maintained. Fire extinguishers are pressure vessels; certain types—mostly older units—have been known to explode or shatter. Because of the number of injuries and deaths attributed to these extinguishers, they should be taken out of service. (Unacceptable and obsolete units are summarized in the selection guide.)

All inverting-type extinguishers should be replaced. These units are no longer manufactured; therefore, suitable replacement parts are not available and these units are no longer listed by UL or approved by FM. The 2½ gal units include soda-acid, foam, cartridge-operated-water, and cartridge-operated loaded-stream types. Of these units, those with brass or copper shells are considered dangerous. When hydrostatically tested, they were several hundred psi below acceptable levels. In one case, a plant replaced all its inverting extinguishers, but decided to discharge the old units before disposal. One unit exploded, killing the operator.

Two other types of units also are considered unsafe: certain dry-chemical units with brass and stainless steel shells and stored-pressure water units with brass or fiberglass shells. The dry-chemical unit shells, which are almost identical to those of the discontinued inverting types, are susceptible to extensive corrosion and had an unacceptably high failure rate during hydrostatic testing. Some incidents of units exploding have been reported. Stored-pressure water types with brass shells are subject to "creep." The bottom of the unit was soft soldered, and may blow out when the unit is used. Fiberglass-shell stored-pressure units tend to rupture upon recharge. However, all of these units have been recalled by the manufacturer; any that may remain should be taken out of service immediately. Withstanding a hydrotest is no assurance that the unit is safe; the test may weaken the fiber structure, causing the unit to explode during use.

Any unit that is improperly cared for can pose a hazard. Carbon dioxide units may explode if subjected to severe neglect and corrosion or if exposed to extremely high temperatures and the relief valve fails to operate. In one incident, a CO_2 unit with internal corrosion exploded while it was hanging on the wall, according to the National Association of Fire Equipment Distributors. The routine 5 year hydrotest had been neglected.

Mixing different types of multipurpose dry chemicals also could cause pressure to build until a unit explodes.

Figure 8-4 Dangers of portable fire extinguishers. Reprinted with permission from *Plant Engineering Magazine*, Technical Publishing, A Company of the Dun & Bradstreet Corporation—all rights reserved.

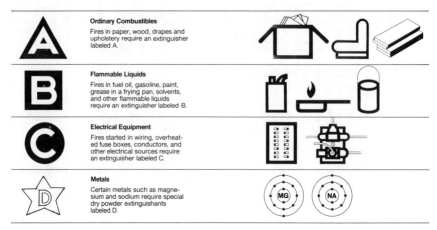

Figure 8-5 Selection and application of portable fire extinguishers. Copyright © National Fire Protection Association. Reprinted with permission.

and women to particular extinguishers, a company can be certain that the right type of extinguisher will be available and that the person who is responsible for it knows how to operate it. One of the latest developments in fire fighting equipment is an all-purpose extinguisher that is effective for Class A, B, and C fires. Its use is still rather limited, however, so the need for instruction in recognizing the various types of fire and correct extinguishers has not been reduced.

Note. Each state regulates its own rules for use of equipment. Be sure to check your own state requirements.

Location and Identification of Equipment

Most extinguishers should be hung or placed so that their tops are not more than 5 feet from the floor. This permits ready removal without injury or accidental discharge.

Heavy extinguishers such as carbon dioxide units of more than 10 pound capacity, water or foam units of more than $2^1/2$ gallon capacity, and so on, should have their tops more than $3^1/2$ feet above the floor.

Extinguisher locations should be marked conspicuously. Signs or arrows should be higher than machines or stock piles for wide visibility.

Do not locate extinguishers in stair wells. Flue actions fans fire and may prevent access to the equipment.

Locate extinguishers close to likely hazards, but not too close. Accessibility and visibility are important considerations.

A Picture Is Worth...

...a thousand words. Manufacturers constantly work to build a better portable fire extinguisher. In the past, refinements have brought improved discharge rates, easier-to-operate squeeze handles, and more effective extinguishing agents. Major changes in portable units today, however, revolve around human engineering instead of hardware engineering.

Most recently, attempts have been made to simplify extinguisher use through pictures. In one modification, picture symbols are being used to identify the class of fire for which a unit is suitable. The pictures are intended to make portable units more effective and safer to use and are replacing or supplementing the color-coded letter shapes that adorn most units. Developed by the National Association of Fire Equipment Distributors, the pictorial label shows all three major fire classes. If the extinguisher is not suitable for a particular class, the picture representing that class is shown in black with a red slash across it. Classes for which the unit is effective are shown in blue. The example (below) would appear on a B and C rated extinguisher—suitable for flammable liquid and electrical equipment fires, and ineffective on fires involving ordinary combustibles.

The pictograph concept is also being used to illustrate how to operate a fire extinguisher and attack a fire (left). Recent tests by an NFPA subcommittee revealed that written operating instructions are often misunderstood. Many people did not know what was meant by the instruction, "pull pin." Others read and followed the recharge guidelines (which appear on the back of a unit) instead of the operating instructions. In subsequent tests, the pictograph instructions were found to be helpful to inexperienced extinguisher users in 36 percent of the test fires that were fought.

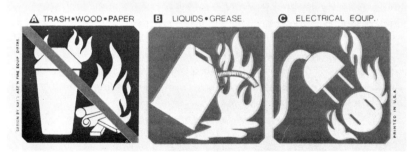

Figure 8-6 Current innovations. Reprinted with permission from *Plant Engineering Magazine*, Technical Publishing, A Company of the Dun & Bradstreet Corporation—all rights reserved.

Avoid placement at locations where extinguishers are hard to reach, or where stock might be piled around them to block access or screen them from view.

Do not place equipment so that it extends out into busy aisles where people, trucks, or dollies might knock it down or damage it.

All extinguishers should be checked once a month to make sure that nozzles are not clogged, that they have been used but not recharged, or that they have not been damaged or tampered with.

All types of extinguisher should be checked and serviced semiannually or annually as directed on the label. All extinguishers should be recharged immediately after using.

Each extinguisher should be tagged to provide a record of its last recharge or inspection date.

GUARDING AGAINST ELECTRICAL FIRES

The biggest single cause of industrial fires is electricity. Most of these fires stem from improper installation of electrical equipment and poor electrical maintenance. Annual property loss of $1.25 billion is due to electrical fires alone, and one factor that contributes to this loss is our tendency to forget that electricity, such a familiar and powerful production ally, is also a destroyer of production facilities. In addition, electricity is largely invisible: wires are generally concealed, and we take it for granted.

EVEN A LIGHT BULB IS A HAZARD

Have you ever seen someone burn his or her fingers changing a hot bulb? The amount of heat generated is pretty close to that of a match, and when a light bulb gets too close to combustible material, a fire can suddenly flash. That's why trouble lamps should have guards to prevent the hot bulb from accidentally brushing against a combustible. Open bulbs in a dusty atmosphere are another hazard. If dust accumulates on a bulb, the dust can char and drop as a spark when the bulb heats up. Prevent a fire by encasing all bulbs in dusty atmospheres in dusttight enclosures.

PROVIDE PROPER PROTECTION

Many times the installation of new equipment inadvertently produces an electrical hazard. Too often, this equipment is powered from old wiring that cannot carry the increased power load without overheating. This eventually leads to a short circuit and a destructive fire.

Never install, or permit others to install, an electrical circuit or wiring system without proper automatic overload protection. Most commonly used devices include fuses, circuit breakers, protective relays, and motor overload relays. They act automatically to de-energize a faulty circuit much as a safety valve depressurizes an overpressured tank. Never overfuse a circuit because there are frequent interruptions caused by a blown fuse or tripped breaker. These are danger signals and should not be ignored. Find the reason for the overload rather than trying to circumvent it. An overfused circuit eventually leads to a fire. Take these five steps to provide proper maintenance for electrical protective devices:

Keep them clean: remove dust, dirt, water, metal filings, or any foreign material.

Tighten bolted, screwed, welded, or soldered connections.

Lubricate to prevent frozen mechanical joints.

Make sure ratings or settings on fuses, circuit breakers, relays, or other protectice devices meet the design specifications of the circuits.

Test protective devices under simulated fault conditions and compare their operations to manufacturer's specifications.

HOW TO HANDLE HAZARDOUS GASES

Almost every production department in almost every industry faces the delicate materials handling problem of inflammable or explosive gases. Here are some tips on the proper storage, handling, and control of hazardous gases:

Dropping cylinders or allowing them to strike against each other may cause a mishap. Hold cylinders in place with chains or other suitable method to prevent toppling.

Oxidizing gases such as oxygen, nitrous oxide, and chlorine must never be stored with the easily oxidized gases such as acetylene, hydrogen, methyl chloride, ammonia, or liquefied petroleum gas.

If cylinders are manifolded, provide adequate hydraulic flash arresters to prevent a flashback through the entire manifold system.

Keep outlet valves on cylinders tightly closed, even when considered empty. Use only tools and wrenches provided by the supplier of the cylinder valves.

Keep electrical equipment in storage areas at a minimum and use only equipment approved by Underwriters Laboratories (UL) for the type of gas stored.

UNDERSTANDING FLASHPOINTS

A *flashpoint* is the lowest temperature at which vapors located above a volatile, combustible substance will ignite in air when the vapor-air mixture is exposed to flame.

Every substance has its own flashpoint, and persons working with such materials should receive training regarding appropriate temperature controls and flame avoidance.

9
OSHA and OSHA Inspection Training

BACKGROUND

The OSHAct includes the following requirements:

Each *employer* shall furnish to his/her employees employment and a place of employment which are free from recognized hazards that are causing or are likely to cause death or serious physical harm to his/her employees.

Each *employer* shall comply with the occupational safety and health standards promulgated under this Act. See Occupatinal Safety Health Standards.

Each *employee* shall comply with occupational safety and health standards and all rules, regulations, and orders issued pursuant to this Act which are applicable to his/her own actions and conduct.

Each *employer* and *employee* comply with the training requirements contained in Part 1910 of the *Code of Federal Regulations* (CFR), commonly referred to as "General Industry" standards. [See Figure 9-1.]

OSHA Inspections

The Department of Labor has hundreds of trained and experienced inspectors who are called safety and health compliance officers (COs). The OSHA COs will visit employers for three reasons:

Death of an employee from work-related causes.

Routine industry inspections in an attempt to imporve the working conditions of the employees in that industry.

Complaints from an employee, if they are in writing, and of sufficient merit to cause OSHA to act.

In any case, an employee representative and management representative have the right to accompany the CO on the tour. The CO should present official government credentials prior to beginning a tour of a firm's facilities.

Compliance Officer's Scope

The OSHAct provides the compliance officer with the following authority and responsibility:

Full access at all reasonable hours.
Must explain why he or she is there.
May request accompaniment on tour by:
   Employee representative.
   Employer's representative.
Is in charge of inspection tour.
May talk with any employee.
Can measure or test noise, air, dust, heat, chemicals, and so on.
May photograph area or location.
Will examine records.
Can request and copy specific medical records and any monitoring data.
Can seek identity of any substances suspected of being dangerous or hazardous. May also take a sample of suspect substance.
May request shutdown, discontinuance of operations, removal of substances, or removal of personnel only when he or she decides there is an imminent danger situation.
Is expected to take proper tests and careful notes.
Must discuss his or her findings with management prior to leaving.
Will advise any complainant whether there was actually a violation.
Can issue a citation for violation.
May recommend a penalty assessment.

RECENT DEVELOPMENTS

Supreme Court Rulings

On May 23, 1978, the United States Supreme Court ruled in the Barlow case that businesses could refuse warrantless OSHA inspections. According to the

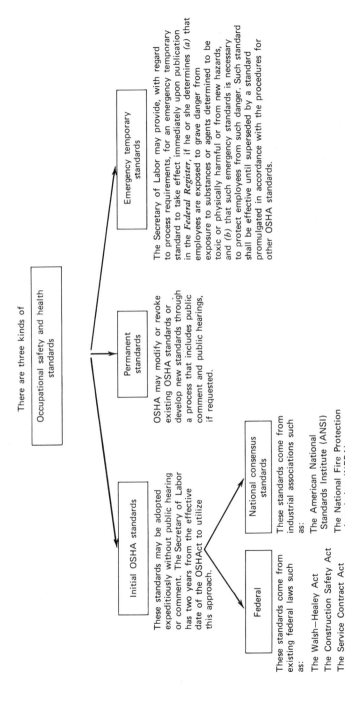

Figure 9-1 Kinds of OSHA standards. From Principles and Practices of Occupational Safety and Health: A Programmed Instruction Course. OSHA 2213, Student Manual Booklet 1, U.S. Department of Labor, Washington, D.C., p. 6a.

Research Institute of America, this decision by the Court put several heretofore unavailable options in the hands of management.

Refusing a warrantless inspection not only forces the OSHA inspector to obtain a warrant, it also limits the scope of his/her search. To obtain that warrant, OSHA must specify what it is looking for. That rules out "wall-to-wall" inspections without probable cause.*

Despite the Court's landmark ruling, only 1.5% of the employers visited by OSHA's compliance officers since the Barlow decision have demanded warrants prior to permitting entrance.†

Challenging OSHA's Authority‡

Before an OSHA inspector arrives, a company should first develop a set of guidelines to use in the following areas:

Determine where the inspector will be permitted to go, with whom, and for how long.

Advise everyone to use a tight-lipped, but cooperative approach, that is, answer all questions honestly, but don't volunteer information not specifically requested.

Before the inspector departs, be sure to ask him or her what areas might receive citations, why, and how to abate them.

Before deciding to appeal (three weeks from the date of the receipt of the OSHA citation), a firm should request an informal conference with the area director. Sometimes in this way differences can be resolved without OSHA issuing a formal complaint.

Any appeal should explain why the company believes the citation is either incorrect or unreasonable (e.g., unreasonable abatement times, excessive fines, unavailable technology).

Any appeal should present an affirmative defense: a faulty search warrant, impossibility of compliance (without reference to cost), unfeasibility of controls, or perhaps proof that the cited violation was an isolated event.

Paying fines just to avoid the expensive and time-consuming legal process can backfire and open a company up to possible civil liability suits.

---

*Industry Week, October 29, 1978.
†Wall Street Journal, July 17, 1978.
‡Industry Week, November 13, 1978.

In October 1978 the U.S. District Court, Southern District of Ohio, ruled that the National Institute of Occupational Safety and Health (NIOSH) is entitled to medical records and subpoenaed from an employer. At the same time the district court ruled that such information should not identify employees by name, since this would constitute an invasion of the employees' privacy. These rulings are expected to be appealed to the Supreme Court.

New Twists to OSHA

OSHA Directive 200-82 instructs its field offices to afford union or employee representatives the opportunity to participate in the opening and closing conferences of an inspection and in any informal conferences between OSHA and businesses where disputes are discussed. The directive also states that any withdrawn or amended violation can be appealed by the union or employee representative to one of the agency's 10 regional offices. This is probably superfluous, since OSHA field office directors have received orders not to withdraw or amend a citation without union or employee approval.*

More than 900 outdated "safety standards" were eliminated by OSHA in October 1978. Included among the deletions are the requirement for split toilet seats and certain specific measurements.†

President Carter's task force on safety and health has issued its First Recommendations Report. It includes the suggestion that OSHA petition the Securities and Exchange Commission to require publicly held corporations to report accident and injury statistics to their shareholders "where a firm's rates significantly exceed industry averages."‡

TRAINING RESPONSIBILITIES

Two broad provisions of the OSHAct are applicable to every employer:

> Every employer must be able to produce records indicating that each employee has received training in the areas identified in the standards.
>
> Every employer must be able to show that each imployee has received training based on an analysis of tasks performed.

---

*Industry Week, October 16, 1978.
†Industry Week, October 16, 1978.
‡Industry Week, October 2, 1978.

Analysis of Requirements

It is easily understood that employers must be able "to produce records" regarding employee training, but what is meant by the term "in the areas identified in the standards?"

OSHA 2254 specifically sets forth all the sections of the act relating to training requirements. These are the "areas identified in the standards." A review of these standards shows them, for the most part, to be directed to particularly hazardous jobs or industries. Some employers might conclude that their operations (i.e., retailing, general office, light manufacturing) are of such a nature as to be excluded from this requirement. However it is important to note that there are three "areas identified in the standards" that are applicable to every employer. These are as follows:

Fire protection.
First aid.
General safety and health (CFR Part 1926.61).

HAZARD ABATEMENT AND APPEAL PROCEDURES

Citations must be posted near the site of violation. Posting is to be made on the day of receipt and must remain for three days or until the hazard has been abated, whichever is longer. Notices of appeal may also be posted. Abatement of a cited condition prior to three days means a notice must be posted alongside the citation.

Appeals from citations and proposed penalties may be made, but a citation becomes binding on employers if not contested within 15 working days of receipt. If the employer wishes to contest a citation, he or she must base the appeal on one of the following points:

The citation itself.
The proposed penalty.
The period established by abatement.
Or any combination of these.

IMMINENT DANGER

OSHAct defines "imminent danger" as any condition that would be expected to cause serious injury or death. See Table 9-1.

Table 9-1 Types of OSHA Violations[a]

These are the types of violations which may be cited, and the penalties which may be proposed:

Other Than Serious Violation   A violation that has a direct relationship to job safety and health, but probably would not cause death or serious physical harm. A proposed penalty of up to $1,000 for each violation is discretionary. A penalty for an other than serious violation may be adjusted downward by as much as 80 percent, depending on the employer's good faith (demonstrated efforts to comply with the Act), history of previous violations, and size of business. When the adjusted penalty amounts to less than $50, no penalty is proposed.

Serious Violation   A violation where there is substantial probability that death or serious physical harm could result, and that the employer knew, or should have known, of the hazard. A mandatory proposed penalty ranging from $300 to $1,000 for each violation is assessed. A penalty for a serious violation may be adjusted downward by as much as 80 percent, based on the employer's good faith, history of previous violations, and size of business. The adjusted penalty for a serious violation may be as low as $60.

Imminent Danger   A situation where there is reasonable certainty that a danger exists that can be expected to cause death or serious physical harm immediately, or before the danger can be eliminated through normal enforcement procedures. An imminent danger may be cited and penalized as a serious violation.

Willful Violation   A violation that the employer intentionally and knowingly commits. The employer either knows that what he or she is doing constitutes a violation, or is aware that a hazardous condition existed and made no reasonable effort to eliminate it.

[a] U.S. Dept of Labor, *All About OSHA,* OSHA 2056, Spring 1980, pp. 25-26 (draft).

When a complaint is made that an imminent danger exists, the compliance officer may make a site investigation. If the condition does in fact present an imminent danger, the compliance officer will do the following:

Post a citation.
Notify employer.
Notify employees affected.
Seek a court order to abate the violation. This could be a request to shut down the equipment or stop using the material in question, and/or remove all personnel from the area except those needed to abate the condition.

10

Obstacle Course to Safety Training

IDENTIFICATION OF HOUSEKEEPING VIOLATIONS

What types of accidents can be prevented through conscientious housekeeping? Here are a few of the more common ones:

- Tripping over loose objects on floors, stairs, and platforms.
- Slipping on wet, greasy, or dirty floors.
- Bumping against projecting or misplaced material.
- Puncturing or scratching hands or other parts of the body on protruding nails, hooks, or rods.
- Injuries from falling objects.

Unfortunately there are no shortcuts to good housekeeping. Employees should be reminded daily, and one of the most effective reminders is to set a good example. If supervisors keep their own offices or work areas looking neat and uncluttered, their campaigns will be more effective.

Regular Procedures

Any department will have fewer hazards and will be more efficient if regular procedures for housekeeping are established.

Identification of Housekeeping Violations   143

- Set aside five minutes at the end of the day for each person to clean his or her own work area. Heavier cleaning chores can be handled in longer periods at the end of each week. Also, intensive cleanups and inspections should be scheduled either on a quarterly, midyear, or year-end basis, depending on the conditions in each department.
- One technique is to post weekly cleanup ratings for each employee, calling attention to shortcomings and recognizing those who are conscientious. Also, in monthly safety meetings, make housekeeping a part of the permanent agenda.

Places to Concentrate

Floor scrubbing, painting, vacuuming, waxing, and other heavy cleaning chores are usually done by janitorial or maintenance crews. It is management's responsibility, however, to be sure that they are performed satisfactorily. This still leaves a big housekeeping job for each supervisor and his or her workers. Here are some areas to concentrate on:

- **Work Areas.** Mark them clearly with painted lines. Plan production carefully so excessive amounts of work materials or finished items don't clutter work areas.
- **Machines and Equipment.** Arrange machines, benches, tables, and other equipment to combine maximum production with enough room for safe, uncrowded work stations. Provide racks or containers for tools, jigs, and fixtures. Keep benches and tables in good repair and cleanly painted.
- **Aisles.** Be sure they are marked clearly and are wide enough to allow easy traffic flow. Keep them free of material, finished parts, and scrap.
- **Floors.** Check to see that they are vacuumed and scrubbed regularly. Protect floors by providing sawdust or other absorbent material on which to place oily parts. Have oil and grease spots removed from the floor immediately.
- **Walls and Ceilings.** Be sure they are scrubbed and painted when necessary. Provide racks for hanging supplies and clothes. A bulletin board will keep calenders, posters, and cartoons off the walls.
    - Other checkpoints include the following:
    - Clean, uncluttered stairs.
    - Adequate storage and care of rags and waste.

- Adequate storage and proper handling of oils, solvents, and other flammables.
- Effective cleanup methods or procedures.
- Adequate storage of raw stock.
- Adequate project storage.
- Clean light fixtures and windows.
- Effective tool storage methods.
- Adequate dust and fume removal equipment.

UNSAFE CONDITIONS INVOLVING MECHANICAL OR PHYSICAL FACILITIES

The total working environment must be under constant scrutiny because of changing conditions, new employees, equipment additions and modifications, and so on. The following checklist is presented as a guide to identify potential problems.

Building

Correct ceiling height.
Correct floor type; in acceptable condition.
Adequate illumination.
Adequate plumbing and heating pipes and equipment.
Windows with acceptable opening, closing, and holding devices; protection from breakage.
Acceptable size doors with correct swing and operational quality.
Adequate railings and nonslip treads on stairways and balconies.
Adequate ventilation.
Adequate storage facilities.
Adequate electrical distribution system in good condition.
Effective space allocation.
Adequate personal facilities (restrooms, drinking fountains, washup facilities, etc.).
Efficient traffic flow.
Adequate functional emergency exits.
Effective alarms and communications systems.
Adequate fire prevention and extinguishing devices.
Acceptable interior color scheme.

Acceptable noise absorption factor.
Adequate maintenance and cleanliness.

Machinery and Equipment

Acceptable placement, securing, and clearance.
Clearly marked safety zones.
Adequate nonskid flooring around machines.
Adequate guard devices on all pulleys.
Sharp, secure knives and cutting edges.
Properly maintained and lubricated machines, in good working condition.
Functional, guarded, magnetic-type switches on all major machines.
Properly wired and grounded machines.
Functional hand and portable power tools, in good condition and grounded.
Quality machines adequate to handle the expected work load.
Conspicuously posted safety precautions and rules near each machine.
Guards for all pinch points within 7 feet of the floor.

OTHER OBSTACLES

Flammables

Flammables are substances that will burn rapidly when they reach their ignition temperature in the presence of an oxidizer.

Flammables encompass a broad range of substances from solids such as paper and rags to liquids such as gasoline to gases such as hydrogen.

Operating employees, supervisory personnel, staff, and passersby, must all be constantly cautioned against the unauthorized presence of flammables, especially in designated areas. Poor housekeeping discipline is one of the primary causes of the illegal presence of flammables.

Smokers or other sources of open flames in combination with flammables can become the team that burns buildings down.

Hazardous Arrangements of Equipment

Equipment position is an important consideration for a variety of reasons.

Unplanned equipment arrangement can result in inefficient production and can be a direct cause of lost profits.

Less obvious, but important nonetheless, is the necessity to consider equipment layout to permit the free and unobstructed flow of personnel and materials through a facility.

Poorly arranged equipment has been known to cause accidents that could have been prevented. Lack of sufficient room for free movement by machine operators and mobile equipment, as well as ample space for waste storage, materials awaiting work or movement, and so on, can be the cause of injury to personnel as well as damage to equipment and facilities.

Personnel should be encouraged to identify potential problem areas caused by the hazardous arrangement of equipment.

11

Hazard Inspection Training

CONTINUING AND SCHEDULED SAFETY OBSERVATION TOURS

The Occupational Safety and Health Act of 1970 makes it necessary for every plant to establish some type of self-inspection and hazards correction program.

Organizing and implementing an OSHA compliance program can be a monumental task if not approached systematically. If undertaken with the philosophy that standards compliance can be tailored to a plant, as well as vice versa, instituting a compliance program need not be overwhelming. Compiling standards abstracts will make the job easier.

The first step is to collect and organize needed information, including the federal (OSHA) standards, state and local codes, the National Electrical Code, and applicable fire protection codes. Some type of easy-to-use photographic equipment will also aid the gathering and documenting of data on plant conditions.

When these materials have been gathered, safety inspection checklists (Table 11-1) and safety and health self-inspection reports (Table 11-2) can be started. Here are some tips that can be helpful.

Keep the job as simple as possible and don't try to do it all at once. OSHA publication 2201, "General Industry: OSHA Safety and Health Standards Digest," can be invaluable when organizing the program. It is available from the nearest OSHA office.

Table 11-1 Safety Inspection Checklist

Name of Facility_____ Date of Survey_____

Address _____
               Street   City

_____
        State   Zip Code   Area Code and Telephone

		Yes	Partly	No

A. MANAGEMENT:

1. Has set the safety policy for the facility. ___ ___ ___
2. Exhibits an active interest in the conduct of an effective safety program. ___ ___ ___
3. Budgets funds for long- and short-range safety program items and projects. ___ ___ ___
4. Holds the superintendent and staff responsible for maintaining a safe environment and for providing a meaningful safety program. ___ ___ ___

B. THE SUPERINTENDENT:

1. Exhibits leadership in the facility safety program. ___ ___ ___
2. Holds the staff responsible for an effective safety program. ___ ___ ___
3. Plays an active role in the operation of the executive safety committee. ___ ___ ___
4. Reports to management on the safety climate and problems. ___ ___ ___
5. Requires an accident reporting and records system. ___ ___ ___
6. Has organized an executive safety program. ___ ___ ___

C. THE PLANT MANAGER:

1. Plans and directs an effective safety program. ___ ___ ___
2. Conducts meetings of the executive safety committee, at least quarterly. ___ ___ ___
3. Provides the superintendent with a copy of minutes of all safety committee meetings. ___ ___ ___

Table 11-1 (*Continued*)

		Yes	Partly	No
4.	Requires that employees be furnished, without cost to themselves, all necessary personal protective safety equipment.	___	___	___
5.	Requires that all supplies, materials, and equipment reflect accepted safety requirements.	___	___	___
6.	Requires that the preventive maintenance program be integrated with appropriate safety requirements.	___	___	___
7.	Requires that foremen maintain safe working conditions and practices in the operations under their supervision.	___	___	___
8.	Provides for the safety training of all foremen.	___	___	___
9.	Requires periodic safety inspections of all work areas by foremen.	___	___	___

D. BUILDING:

1. Structurally sound. ___ ___ ___
2. Ramps provided. ___ ___ ___
3. Entrances—one primary entrance for wheelchairs. ___ ___ ___
4. Doors and doorways—adequate width. ___ ___ ___
5. Doors and doorways—landing platforms meet standards. ___ ___ ___
6. Doors and doorways—door sills meet standards. ___ ___ ___
7. Stairs meet standards. ___ ___ ___
8. Steps have rounded nosings. ___ ___ ___
9. Stair handrails are standard. ___ ___ ___
10. Stair handrails extend beyond top and bottom steps 18 inches. ___ ___ ___
11. Stair step design is standard. ___ ___ ___
12. Floors have nonslip surface. ___ ___ ___
13. Floors are free of holes and projections. ___ ___ ___
14. Floors on a given story have common level. ___ ___ ___
15. Toilet rooms are designed for use by handicapped persons. ___ ___ ___
16. Water fountains are designed for use by handicapped persons. ___ ___ ___
17. Public telephone is installed for use by handicapped persons. ___ ___ ___

Table 11-1 (*Continued*)

			Yes	Partly	No
	18.	Elevators are accessible for use of handicapped persons.	___	___	___
	19.	Switches and controls for light, heat, fire alarms, and all similar controls of frequent or essential use are placed within reach of person in wheelchair.	___	___	___
	20.	Facilities in building that would be hazardous for the blind should be specifically identified for recognition by blind persons.	___	___	___
	21.	Visual and audible warning signals are interrelated for benefit of the blind and deaf.	___	___	___
	22.	Exit signs meet standards.	___	___	___
	23.	Aisles are defined with marking lines on floor.	___	___	___
	24.	Aisles are wide enough to accommodate two wheelchairs.	___	___	___
	25.	Protruding hazards dangerous to blind persons are eliminated.	___	___	___
	26.	Physical hazards marked in accordance with safety color code.	___	___	___
E.	GROUNDS:				
	1.	Safe conditions for entering and exiting privately owned vehicles.	___	___	___
	2.	Hazards on facility grounds have been eliminated or controlled.	___	___	___
	3.	Public access roads for employees using public transportation.	___	___	___
	4.	Parking lots used by employees meet standards.	___	___	___
F.	STORAGE:				
	1.	Storage areas and rooms are orderly.	___	___	___
	2.	Stored material are handled safely.	___	___	___
	3.	Mechanical materials handling equipment is safely operated.	___	___	___
G.	MACHINERY:				
	1.	Machinery moving parts are provided with protective guarding.	___	___	___

Table 11-1 (*Continued*)

		Yes	Partly	No
	2. Instruction is given on safe use of power tools.	___	___	___
	3. Supervision follow-up on power tool safety instruction is adequate.	___	___	___
	4. Point of operation guards have been installed.	___	___	___
	5. Point of operation guards are used by personnel.	___	___	___
H.	LIGHTING:			
	1. Illumination levels for seeing tasks meet standards.	___	___	___
	2. Illumination sources in hazardous areas meet standards.	___	___	___
	3. Illumination on ramps and stairways is satisfactory.	___	___	___
I.	VENTILATION:			
	1. General ventilation of work areas is satisfactory.	___	___	___
	2. Hazardous fumes are removed to safe discharge points.	___	___	___
J.	PERSONAL PROTECTIVE SAFETY EQUIPMENT:			
	1. Proper types of equipment are provided by management.	___	___	___
	2. Supervision follow-up ensures use by personnel.	___	___	___
K.	COMPRESSED GAS CYLINDERS:			
	1. Storage and handling is in accordance with safety standards.	___	___	___
L.	ELECTRICAL WIRING SYSTEM (PERMANENT AND TEMPORARY)[a]			
	1. Wiring meets needs of the facility.	___	___	___
	2. Sufficient number of convenience outlets.	___	___	___
	3. Overloading of circuits is controlled.	___	___	___
	4. Use of extension cords is controlled.	___	___	___
	5. Emergency switches are spaced and functioning satisfactorily.	___	___	___

Table 11-1 (*Continued*)

		Yes	Partly	No

M. ELEVATORS:

    1. Elevators are inspected and maintained on a periodic and scheduled basis. ___ ___ ___
    2. Elevator cars and shaftway openings are protected by doors or gates interlocked with elevator controls. ___ ___ ___

N. PRESSURE VESSELS:

    1. Boilers are inspected and approved by competent inspectors at least annually. ___ ___ ___
    2. Air receivers are inspected and approved by competent inspectors at least annually and condensate is drained regularly. ___ ___ ___

O. ELECTRICAL EQUIPMENT:

    1. Equipment is correctly fused. ___ ___ ___
    2. Equipment is properly maintained and cleaned. ___ ___ ___
    3. Portable electrical equipment is grounded. ___ ___ ___

P. FIRE HAZARDS:

    1. Housekeeping is generally acceptable. ___ ___ ___
    2. Housekeeping in above-normal risk areas is adequate ___ ___ ___
    3. Waste containers are adequate in design and availability. ___ ___ ___
    4. Waste containers are properly disposed of at end of each day. ___ ___ ___
    5. Waste paper and rags are disposed of regularly. ___ ___ ___
    6. "No Smoking" signs are posted and observed. ___ ___ ___
    7. Flammable liquids are safely dispensed. ___ ___ ___
    8. Flammable material is stored safely. ___ ___ ___
    9. Wiring in hazardous areas meets National Electrical Code. ___ ___ ___
    10. Spray painting booths meet standards. ___ ___ ___
    11. Spray painting areas are kept free of accumulations. ___ ___ ___

Table 11-1 (*Continued*)

		Yes	Partly	No

Q. FIRE PREVENTION CONTROLS:

   1. Fire extinguishers are properly serviced and maintained.
   2. Personnel periodically are given instruction in the use of extinguishers and fire hose.
   3. Fire instructions for each floor have been issued and discussed with staff and clients.
   4. Sprinkler heads are unobstructed.
   5. Fire doors are functional.
   6. Fire escape stairway is free of obstruction.
   7. Fire aisles are kept clear.

R. MEDICAL PROGRAM:

   1. Doctor on duty and in charge of medical program.
   2. Registered nurse on duty.
   3. First aid room and facilities available on premises.
   4. Contract medical services provided (alternative to items 1 and 2).

S. PERSONAL PROTECTIVE EQUIPMENT:

   1. Each person exposed to eye hazards is furnished with personal goggles or a face shield that meets standards.
   2. Each person exposed to respiratory hazards is furnished with a personal respirator that meets standards.

T. SPECIAL ITEMS:

Note (Add special items if desired.)

Name (s) and Title (s) of Person (s) Involved in Survey:

_____

_____

---

[a]Must conform to the requirements of the National Electrical Code.

Table 11-2 Safety and Health Self-Inspection Report[a]

Area Inspected_____ Date_____ Inspector_____

Code	Hazards	Corrective Actions

Analysis and Comments

[a] Principles and Practices of Occupational Safety and Health: A Programmed Instruction Course, OSHA 2215; Student Manual Booklet 3, U.S. Department of Labor, Washington, D.C., p. 45.

Plan on many once-over-lightly inspections with later in-depth follow-ups. See list of possible problems to be inspected (Table 11-3).

HIGH ACCIDENT AREAS AND OPERATIONS EMPHASIZED

Proceed on a "worst first" basis. There are two reasons for doing this: any serious or fatal accident must be immediately reported to OSHA and may trigger an official inspection; and the likelihood of employee complaints will be reduced.

In terms of cost effectiveness, far more accidents will be prevented (therefore, dollars saved), by concentrating attention on the worst accidents first. Analysis of the company's accident history should reveal where and how

Table 11-3 List of Possible Problems to Be Inspected[a]

Acids	Hoists
Aisles	Hoses
Alarms	Hydrants
Atmosphere	Ladders
Automobiles	Lathes
Barrels	Lights
Bins	Mills
Blinker lights	Mists
Boilers	Motorized carts
Borers	Piping
Buggies	Pits
Buildings	Platforms
Cabinets	Power tools
Cables	Presses
Carboys	Racks
Catwalks	Railroad cars
Caustics	Ramps
Chemicals	Raw materials
Claxons	Respirators
Closets	Roads
Connectors	Roofs
Containers	Safety devices
Controls	Safety glasses
Conveyors	Safety shoes
Cranes	Scaffolds
Crossing lights	Shafts
Cutters	Shapers
Docks	Shelves
Doors	Sirens
Dusts	Slings
Electric motors	Solvents
Elevators	Sprays
Explosives	Sprinkler systems
Extinguishers	Stairs
Flammables	Steam engines
Floors	Sumps
Fork lifts	Switches

Table 11-3 (*Continued*)

Fumes	Tanks
Gas cylinders	Trucks
Gas engines	Vats
Gases	Walkways
Hand tools	Walls
Hard hats	Warning devices

*"Principles and Practices of Occupational Safety and Health: A Programmed Instruction Course," OSHA 2213, Student Manual Booklet 1, U.S. Department of Labor, Washington, D.C., p. 40.

and when its most costly accidents are occurring. This could be an occasional accident that is costly in terms of death, disability, and/or many lost person-days, or it could be an inexpensive accident that rarely results in more than one or two lost person-days but occurs frequently.

Whether it is a hazardous area where accidents are taking place or a hazardous operation that is the site of costly accidents, the company will realize greater return on its investment in safety training by concentrating its attention on the "worst first."

PLANNING A FORMAL INSPECTION FOR SAFETY AND HEALTH HAZARDS

Objectives of Planning a Formal Inspection

A formal safety and health program has two main objectives:

- To recognize all undesired practices and conditions that could result in personal injury or the impaired health of the employees, or damage to equipment, and to identify relevant standards.
- To evaluate and record the general degree of hazard associated with each item as a guide to setting up remedial and preventive action priorities.

Three Principles in a Planned Safety and Health Inspection

When performing a planned inspection, keep in mind the following principles:

Planning a Formal Inspection for Safety and Health Hazards

OSHA standards require inspections in certain situations.

Normal wear and tear eventually brings trouble.

People make mistakes.

The regularly scheduled inspections should come at planned times during the year, month, week, or day, depending on the particular circumstances, or on the item(s) to be inspected. For example, it is generally a good idea to check for cleared work areas (good housekeeping) daily, whereas fire extinguishers are usually checked monthly. Therefore it is necessary to decide on when, or at what times, certain items need to be inspected.

Poor Planning Leads to Poor Results

- The planning of a formal inspection is very important.
- If planning is poor, results will also be poor.

Inspections Should Be Well Organized

To be effective, any inspection program must be well organized and planned in a way that assures complete and timely inspection of all facilities. This means having regular, systematic inspections, in addition to the ongoing ones that always should be in process (i.e., constantly looking for hazards during the daily routine.)

Daily inspection alone, even though organized, planned, and embodying a checklist, will not identify *all* the hazards in a work area. Periodic, planned inspections will be most beneficial in identifying additional hazards, as well as helping to decide what steps must be taken to control them.

Questions to Answer Before Performing an Inspection

To effectively plan for a formal inspection, management should always require that written answers to the following key questions be submitted prior to the inspection:

- What problem will be inspected?
- What are the critical factors for each problem to be inspected?
- What conditions will be inspected?

Two Types of Planned Safety and Health Inspection

There are two types of planned inspection that a supervisor might use to good advantage:

Inspection of the supervisor's own work area.

"Audit" inspection of another supervisor's work area.

The first type of inspection is vitally important because the employer is responsible for what happens in that work area, and the supervisor's position makes him or her a direct representative of management. Also, the supervisor is probably the person most familiar with all aspects of the work area and its relationship to the elements of hazard analysis previously discussed.

Audit Inspections Help Everyone

Audit inspections are planned in the same manner as inspection of one's own work area, which is described later. The only difference is that in an "audit" inspection one supervisor inspects another supervisor's work area. An important point to keep in mind when performing an audit inspection is that the goal is to HELP the other supervisor by discovering problems he or she may have missed.

Benefits of an Audit Inspection

Audit inspections produce a two-way benefit. First, they help two supervisors to do each other and their employees a good turn. Second, they help each supervisor to gain a new perspective. After looking over someone else's work area, a supervisor may come back to his or her own area with added or better ideas for improving that operation. As can be seen, there ARE benefits to be gained by performing audit inspections.

Objectivity Is Important

The importance of the second type of inspection, "audit inspection of another work area," may not be as immediately obvious as the first type. Oddly enough, one of the greatest advantages a supervisor has in a facility inspection—familiarity with the employees, equipment, and machines in that area—can be a disadvantage. Personal involvement with a work area can rob a supervisor of one of the most critical ingredients of a good safety and health inspection, namely, objectivity. The less objective a supervisor is, the more likely it is that he or she will miss dangerous hazards.

Being human, people have trouble remaining totally objective if they are personally involved in a situation. Almost all people tend to lose some degree of objectivity in repeatedly inspecting their own work area. One factor is the tendency to overlook anything not included on the checklist; thus the checklist can become a disadvantage as well. Checklists are guides and should never be considered as final, all-inclusive, or never-changing.

Hazards Tend to Accumulate

Many safety or health problems tend to accumulate very gradually. This gradual change is hard to recognize if the same person who does the inspection works day after day in the same area.

In an audit inspection, the cumulative effects of any gradual change will be more noticeable to a supervisor from another work area.

Formal Inspections

For purposes of discussion, the audit inspection and the planned, periodic inspection are termed *formal* inspections. They occur at certain times and, as will be seen later, they give a great amount of detailed and useful information. Since formal inspections are usually the most thorough, they are very important. Remember, however, that these formal inspections cannot replace daily inspections by supervisors.

Note. One point to keep in mind is that each employee should fix the hazards that he or she has the authority to correct. Don't wait until someone is injured or becomes ill before exercising authority that has already been given.

OSHA STANDARDS USED AS INSPECTION GUIDES

Begin to organize the checklists by referring to the first page of the standards—actually, a table of contents. Note that some of the subparts may not apply to all plants and can, therefore, be ignored in certain cases. Become familiar with, and evaluate, the individual subparts. For example, consider the following:

> **Subpart A—General.** This section applies to all plants but contains no direct references to actual plant requirements. It is worth mentioning that some of the general standards widely publicized as "scare" items have no basis in fact. Rather, they have been fabricated by opponents of OSHA and the OSHAct. Sections 1910.3 and 1910.4 outline opportunities available if a standard can be documented as unworkable.
>
> **Subpart B—Adoption and Extension of Established Federal Standards.** If any of these standards apply to your plant, obtain copies of the applicable standards.
>
> **Subpart R—Special Industries.** Plants not within the industries listed are not obligated to comply with any of these standards. However some of them might be considered useful for other industries establishing a very thorough compliance program.

Once the standards have been generally studied, they must be examined in detail. With a transparent yellow marker, check the standards that specifically apply and will require immediate action. (The yellow marks will not show on photo copies that are made later.) It is also a good idea to remove the staples from one set of the OSHA standards and keep it in a loose-leaf binder so that pages can be easily removed, copied, and replaced.

Next, make a copy of each page on which standards have been marked for attention, cut out the needed paragraphs, and glue them on the left-hand side of a blank sheet of paper. This arrangement leaves room for note making. Cutting out paragraphs and remounting them permits editing of the unnecessary material and the bringing together of related standards that may not appear consecutively in the original publication. The finished OSHA standards abstract (Figure 11-1) is now ready to be copied and distributed as necessary.

A copy of each standards abstract, including any instructional notes, should be retained in follow-up files. These files can be arranged by department involved or by standard. It may be advantageous to use both systems with cross-referencing.

Using an abstract system to organize OSHA compliance paperwork offers many advantages. An abstract does the following:

- Helps direct attention and establish priorities.
- Allows various related standards to be combined.
- Legally justifies action to be taken by reproducing the exact wording of standards.
- Educates personnel who receive it.
- Provides basic documentation for the compliance program.
- Reduces time needed to explain the why's and how's of a compliance requirement.

SIMULATED PLANT INSPECTION

Now let's accompany a hypothetical OSHA compliance officer on a walk-around inspection of a workplace. This is an opportunity to review some potentially hazardous conditions that are found in many workplaces all over the United States. As the simulated inspection is conducted, we will attempt to create a mind's-eye picture of the situation, describe the hazard(s), and consider the various alternatives to abating each of them.

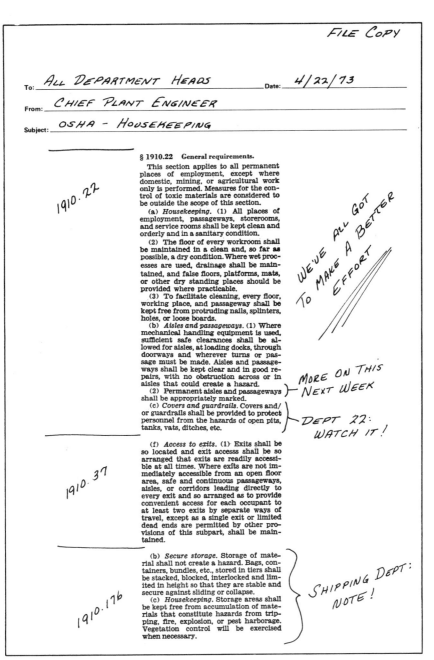

Figure 11-1 OSHA standards abstract memo. Reprinted with permission from *Plant Engineering Magazine*, Technical Publishing, A Company of the Dun & Bradstreet Corporation—all rights reserved.

Situation 1

Perception: An employee, probably assigned to the maintenance department, is using a portable piece of cleaning equipment to work on the floor of a manufacturing area.

Hazard: Close-up inspection reveals that one of the mechanical guards is missing from the equipment. Additionally, our sound level meter indicates a noise problem.

Abatement: Locate or fabricate the missing guard and request that it be installed. Instruction regarding future actions to keep guards in place may be needed as well. An attempt should be made to muffle the noise sufficiently to bring it into compliance with legal limits. If this cannot be accomplished, it may be necessary to replace the equipment with a model that does not offer this noise hazard.

Situation 2

Perception: An employee is climbing a portable wooden ladder in a backroom storage area.

Hazard: There are no railings on the landing platform and the toeboard is missing. The ladder is too short at the point of access to the landing (it should extend 36 inches). The employee is facing the ladder as he should; however he is observed to be supporting himself with only one hand. The other hand appears to be used for carrying tools or materials.

Abatement: The department foreman should be advised of the physical hazard (i.e., the missing toeboard), with a recommendation that this requirement, as well as the ladder and climbing procedures, be a topic for continued inclusion in training sessions.

Situation 3

Perception: In checking the floor areas, an open grease pit in the company garage is noted.

Hazard: Although sockets for guard rails exist, railings are not in use to inhibit personnel from falling into the pit. Additionally, we find an oily rag on the floor and generally poorhousekeeping.

Abatement:	The rails should be located and installed, and the area cleaned up. This appears to be a point for regular follow-up during safety meetings.

Situation 4

Perception:	We begin to walk down a stairwell and note a 55-gallon drum at the base of the stairs.
Hazard:	The stairway, therefore personnel passage, is partially blocked by the drum.
Abatement:	The drum should be removed and personnel reminded to keep possible fire exits clear of obstructions AT ALL TIMES.

Situation 5

Perception:	As we continue down the next flight of stairs, a large pedestal floor fan blocks our way at the base of the stairs.
Hazard:	Same as Situation 4.
Abatement:	Same as Situation 4.

Situation 6

Perception:	As we walk by an area where carpenters are remodeling a suite of offices, we notice a large hole in the floor.
Hazard:	The hole should be safely covered and/or surrounded by temporary barriers, with appropriate signs to reduce the likelihood of someone tripping or falling into or through the hole.
Abatement:	This is another reminder item for inclusion in safety meetings.

Situation 7

Perception:	We step out onto the top level of the building. The equipment and ventilators on the roof have to be maintained regularly, so there is frequent need for employees to be here. The guard rail at the perimeter is good, however poor housekeeping (see Hazard) could cause its strength to be put to the test.

Hazard: Many items have been carelessly strewn about this level and offer a real tripping/falling hazard.

Abatement: The area should be cleaned up, and all serviceable items should be placed in the appropriate storage areas. Unserviceable materials should be carefully stored to await repair or appropriate disposition. This item should be a point for discussion at safety meetings.

Situation 8

Perception: As we enter a large production area with many machines (cranes, drills, saws, etc.), we recognize the presence of oily, greasy floors.

Hazard: Oily or greasy floors constitute hazards that bring on falls, and falls cause injuries ranging from fractures to twisted backs or even worse. Additionally, fire potential is significantly increased with oily floors.

Abatement: The walking areas should be degreased or covered over immediately and personnel traffic removed until the hazard is eliminated. Operators and maintenance personnel should be made aware of their responsibilities in keeping this hazard to a minimum.

Situation 9

Perception: As we continue our inspection we note a wheel-mounted fire extinguisher positioned in its floor-marked storage area.

Hazard: A 55-gallon drum has been carelessly left in front of the extinguisher storage area blocking access to it. In addition, the area is not clearly marked so that personnel can locate it rapidly in an emergency.

Abatement: The area should be specifically identified as containing a fire extinguisher. All personnel should be cautioned against blocking access to any emergency equipment.

Situation 10

Perception: Next we see a wall-mounted fire extinguisher posted on the inside of a large door jamb.

Hazard: Closer observation reveals that a bundle of welding rods lies on the floor below the extinguisher. A spectacular spill could result if someone ran up to grab the extinguisher and slipped on those rods. Of course this would happen only if the extinguisher could be found and there are no wall markings on either side of the door jamb to call attention to its presence.

Abatement: Maintenance personnel should be regularly reminded of their safety-related responsibilities, and in this situation they can be shown that they haven't been discharging their responsibilities. This applies to both job cleanup and fire extinguisher placement and wall markings.

Situation 11

Perception: As we walk through one of the ground floor exits, we find a storage area for compressed gas cylinders. Some are sitting on their base and secured by a chain, but most are either leaning against a wall or some less secure support or are flat on the ground.

Hazard: Compressed gas cylinders must be stored and used safely. All cylinders should be identified and segregated according to whether they are full or empty. Proper storage requirements must be implemented.

Abatement: Personnel with the responsibility for caring for this material should be advised of the consequences of neglecting storage requirements. Sufficient areas should be made available to allow for meeting these requirements.

Situation 12

Perception: Next we approach a specific metal cylinder used to store compressed gases.

Hazard: Close-up inspection reveals an unknown substance that seems to have come from inside and is surrounding the cylinder cap.

Abatement: Isolate the storage area and do not move the cylinder. Contact the supplier to identify the level of hazard and the recommended method of elimination.

Situation 13

Perception: As our walk-around inspection continues, we see a maintenance crew storage and work room that is immediately dubbed "the snakepit."

Hazard:	What we've found is a room that seems to exemplify a careless attitude toward good housekeeping and safe practices on the part of the people who use this area. We see tools strewn on raised work surfaces as well as the floor. Additionally, we see hundreds of feet of heavy electrical cord and rope, some tied in loose bundles, haphazardly located throughout the area. Some power drills with bits inserted are plugged in and resting on the floor near the base of the raised working surfaces.
Abatement:	It isn't enough to settle for cleaning this area up. The appearance of this room is a clear indicator of a need for additional safety training as well as a better attitude about responsibilities for safety. This one will take awhile.

Situation 14

Perception:	Next we find an electric motor that drives a machine through a mechanical power transmitter.
Hazard:	Here we have a chain drive transmission exposed without any form of guarding.
Abatement:	Locate and reinstall the original guard. If it cannot be found, one should be fabricated to provide necessary protection from fast-moving pinch points.

Situation 15

Perception:	A look outside the plant reveals a parked truck on an inclined driveway adjacent to the dock.
Hazard:	If the truck begins to move while personnel are working in or near it, there is a high probability of injury or death.
Abatement:	Vehicles parked in a dock area should be chocked. Drivers and materials handlers should be so advised, and a warning sign should be posted as a reminder if there is not one there already.

Situation 16

Perception:	In the production area we see a medium to high speed production line machine that could cause its operator to lose his or her fingers or entire hand.
Hazard:	We see that the guarding system does not sufficiently inhibit

	the machine's operator from getting the hands too close to the machine while it is operating.
Abatement:	The machine requires immediate guarding. Until such guarding is provided, however, operators of the machine should not be expected to expose themselves to possible injury. As a part of their safety training, all personnel should receive frequent reminders to call situations like this to management's attention so that hazards can be expeditiously eliminated.

Situation 17

Perception:	One wall has been established as a storage point for hanging slings that are not currently in use but are always on call.
Hazard:	At first glance the situation looks acceptable. Unfortunately, a closer look reveals that most of the slings are kinked and frayed, thus exposing their users to accidents due to broken slings and dropped loads.
Abatement:	A regular inspection and maintenance program should be established to eliminate the high risks associated with a kinked or frayed sling.

So far we have seen some cases of standards related to walking and working surfaces, guarding, fire protection, housekeeping, and other areas.

Now let's introduce another ingredient—occupational health problems. This is an area that at last is receiving the long overdue recognition and attention it deserves as a vital component of the total safety and health system.

Many, if not most, work situations have actual or potential health hazards. Some of these are illustrated in the continuing series of situations.

Situation 18

Perception:	Over in the plating area, where employees are regularly exposed to strong acids as a part of their daily operations, we see an eyewash fountain.
Hazard:	As a precaution, we check the operation of the fountain by pulling the chain, which should cause the water to flow freely. Much to our dismay, we note that this emergency device fails to function. In the event of an emergency, an employee who had immediate need of the eyewash fountain would be out of luck.

Abatement: A regular maintenance and checkout program should be established to ensure the immediate availability of this equipment in the event of an emergency.

Situation 19

Perception: Over in the locker room we find an industrial sink. Adequate washing facilities are, we recall, also important for controlling skin diseases. Dermatitis is a common type of occupational illness.

Hazard: Stepping on the operation ring near the floor to verify the utility of the sink, we see that only a trickle of water is coming out of the pipes.

Abatement: A regular maintenance and operational checkout program should be established to ensure that washing capability is always available.

Situation 20

Perception: Moving through the welding area we note the presence of arc welding.

Hazard: Arc welding creates exposure to ultraviolet light, which may result in eye damage. It similarly involves exposure to air contaminants, since welding on mild steel generates iron oxide fumes. When the fumes pass in and around the welder's face shield, he or she may breathe in harmful amounts.

Abatement: A local exhaust system should be installed to effectively remove fumes resulting from arc welding. If the welder is properly positioned (i.e., out of line of the main fume draft), exposure is greatly reduced and general workplace ventilation may be sufficient to adequately control exposure.

Situation 21

Perception: Even before we enter a punch press area and see the signs identifying it as a high noise area, we can hear the tremendous sounds of the impact of metal on metal.

Hazard: At the outset we are not aware of what effort has already gone into attempting to eliminate or reduce the noise problem. However several employees are working in the area, and all but one are wearing personal ear protection.

Simulated Plant Inspection

Abatement: First, a scientifically controlled analysis of the noise intensity in the area should be conducted. Assuming there is a problem, an attempt should be made to physically eliminate or reduce it. Effective engineering control is considered to be the only real solution to excessive noise problems. The use of personal ear protection should be considered to be a temporary expedient for use only when engineering or administrative controls cannot solve the problem. Furthermore, it appears at first glance that not all the employees consider the excessive noise level to be a problem. This is a topic that should receive regular attention as a part of the safety training program.

Additionally, the regular measurement of the noise exposure in an operation where the noise may be hazardous to hearing should be incorporated into the overall safety program. The use of either ear muffs or ear plugs for personal protection should be an individual's own choice. Together, they may reduce noise exposure as much as 30 dBA (decibels on the "A" scale of a noise measuring device).

Accomplishing this degree of attenuation, however, depends to a great extent on the willingness of the employee to undergo some degree of physical discomfort involved to achieve the protection. In cases such as this, the exact attenuation of noise exposure cannot be estimated exactly; however its effectiveness can be measured audiometrically.

Now let's continue our simulated walk-around inspection on a construction site.

Situation 22

Perception: As we enter the site, we see two construction workers quietly talking.

Hazard: A much closer look reveals that one of the two "hard hats" is holding a portable electric radial saw at his side and that the saw has no guarding. Until the saw is put to use, no real hazard exists unless it is accidently dropped on a foot. When the saw is turned on, however, its operator will be immediately and unnecessarily exposed to the rapidly turning blade.

Abatement: The specially designed guard that came with this saw should be reattached before the saw is placed in use. A quick check of other saws and assorted power-operated equipment is worthwhile to identify whether such laxity is an isolated instance or a practice that pervades the entire construction site. This topic should be mentioned regularly as part of the safety training program.

Situation 23

Perception: At the top of the building under construction an open elevator shaft is noted—eight floors up.

Hazard: Any employee or inspector could trip and fall to his or her death for lack of some physical barrier to keep people away from the shaft.

Abatement: Both a physical barrier and warning signs should be placed in proximity to this hazard. Employees should receive instructions regarding the necessity for isolating hazards such as this.

Situation 24

Perception: Walking through the storage and staging area, we note a dramatic example of a real problem that exists not only in construction sites but in industrial plants, warehouses, and even retail operations—namely, a stack of boxes that could be a model for the Leaning Tower of Pisa.

Hazard: The boxes are stacked too high, there are defective pallets, and the ground is so soft that inevitably the stack will come crashing down. We can only hope that no one will be near it when this occurs.

Abatement: Before anything else happens, the area surrounding the leaning stack of boxes should be posted as "off limits" to nonessential personnel. Then appropriate heavy equipment should be used to restack the boxes, using dependable pallets. When the emergency situation has been corrected, the long-term implications of such careless materials handling should be addressed.

As part of their safety training, personnel responsible for stacking operations should receive regular reminders of the importance of caution and care in the performance of their tasks. Other personnel, whose responsibilities take them to areas where materials are stacked, should be made aware of the necessity to stay alert to problems of this type and to report their existence right away.

Situation 25

Perception: We can't believe our eyes. A "hard hat" worker is walking directly beneath a load of plywood being transported by an overhead crane.

Hazard: Hard hat or no hard hat—the worst place a worker could possibly be is directly beneath an airborne load of anything, but this case is especially dangerous, since we see that only a single choker sling is in use.

Abatement: Assuming that the worker is unaware of the hazardous situation he has walked himself into, he should be warned to move away from the path of the overhead load. Following the abatement of the immediate threat to the worker, a series of cautionary remarks should be included as a part of the safety training program. Make sure that crane operators get the word, too.

Situation 26

Perception: As we continue to observe the construction activities, we note two men operating air hammers while a third is clearing away broken pieces of concrete with a shovel. None wears eye, ear, or foot protection.

Hazard: None of the three has the necessary personal protection to ensure his individual safety and health. Excessive noise levels generated by the air hammers will, over a substantial exposure period, cause all three to lose some degree of their hearing capacity. All three are exposed to flying and falling fragments of the concrete, as well, and they should have eye and foot protection.

Abatement: A program designed to increase desire to use personal protective equipment should be instituted. The desire is just as important as knowing how and when to use the equipment. No supervisor can keep an eye on all employees all the time, and if motivation to use the equipment doesn't exist, it won't be used or maintained properly.

Situation 27

Perception: Looking inside the mobile workshop, we see a bench grinder.

Hazard: There are no spark-deflecting guards on either the abrasive wheel or the wire wheel, and there is no face and eye protec-

	tion visible. Additionally, the abrasive wheel is worn and needs to be replaced.
Abatement:	Once the immediate hazards have been corrected, appropriate personnel should receive periodic reminders to be alert to the need to correct these hazards whenever they occur.

Situation 28

Perception:	Also inside the mobile workshop is a floor-mounted buzz saw used to split lumber.
Hazard:	As in situation 22, personnel who operate the saw as it sits will be exposed to the dangers of an unguarded, rapidly spinning saw blade.
Abatement:	The guard should be located and reinstalled on the saw before it is used again. A reminder sign should be posted to let saw operators know of the requirement to use the saw only when the guard is in place. Regular reference to this requirement should be included in safety training meetings.

Situation 29

Perception:	Moving back outside we note that three levels of scaffolding are in use in constructing this brick-facing building.
Hazard:	Numerous hazards exist in this situation, for example, inadequate cross-bracing on the uprights of the scaffolding, use of poor materials such as warped planks, and to top off this collection—there are no guard rails or toeboards. Just a moment, we note still another hazard—the center upright at the base of the wooden scaffolding is almost unbelievably insecure.
Abatement:	Stop the entire operation and correct all the noted deficiencies before continuing the construction of this building. The person in charge of this site should be severely reprimanded or even dismissed depending on his or her previous record. All future training should make note of the necessity to adhere to these and all safety-related requirements.

Situation 30

Perception:	Passing by a generator used to power some electrical equipment, we stop to review the situation.

Hazard:	The belt and pulley require guarding inasmuch as the starter button is so close that the operator must reach into the danger area to activate the generator.
Abatement:	The guard for this part of the generator should be located and put into use prior to further operation. Personnel need to be cautioned not to forget to replace the guard after it has been removed for maintenance.

Situation 31

Perception:	At the far end of the site we can see a long-necked crane (aerial boom) in operation.
Hazard:	Of all the dangers involved in this type of operation, the use of this equipment in unreasonable proximity to overhead electrical lines is one of the most serious. Construction standards set forth specific operating clearances with respect to energized lines.
Abatement:	As a regular part of the safety training program, all personnel working with or on cranes should be made aware of the possible results of operating a crane too close to loaded electrical power lines. Furthermore, the company should require that only insulated, approved cabs be used when personnel are needed at the working end of a boom.

Situation 32

Perception:	As we turn to leave the inspection site, a chain is spotted hanging on its storage rack.
Hazard:	A closer look reminds us of the old story about a chain being only as strong as its weakest link. Here we see several links that appear to be worn and elongated—a characteristic of overloading. Also visible is evidence of improper splicing. So how strong is this chain?
Abatement:	A regularly scheduled program of chain (and sling) inspection, maintenance, and replacement is needed. Using personnel should be trained to spot irregularities and to take appropriate action.

This view of some situations you and a compliance officer might see in the course of visiting workplaces was designed to provide insight into some safety and health hazards and their analysis. This is the type of hazard recognition training employees need to help them get started. It is never too late to begin.

12

Medical and First Aid Training

PROGRAM OBJECTIVES

The health and safety of employees must receive top priority in every business. Because professional medical help is seldom immediately available, it is imperative that emergency treatment be available when an injury occurs.

Occupational health programs are concerned with all aspects of a worker's health. The basic objectives of such a program, according to the American Medical Association, should be the following:

- To protect employees against health hazards in their work environment.
- To facilitate job placement and to ensure the suitability of individuals according to their physical capacities, mental abilities, and emotional makeup without danger to their health and safety or to that of their fellow employees.
- To assure adequate medical care and rehabilitation of those injured or made ill on the job.
- To encourage personal health.

The American Medical Association has listed the following elements and services that should be provided to maintain the health of every work force, regardless of size of the operation:

- Maintenance of a healthful environment.
- Diagnosis and treatment.
- Health examinations.
- Immunization programs.
- Medical records.
- Health education and counseling.
- Free communication between the plant physician and the personal physician.

Achieving these goals benefits both employee and employer. If employees are to perform the jobs safely and efficiently, they must be in good health.

PERSONNEL SELECTION AND TRAINING

Assigning the right person to the right job and protecting him or her against the hazards of the work are, however, only the beginning. Medical personnel today must be familiar with the on-the-job hazards, and they must continually assess the job hazards just as they periodically reexamine employees to assure their physical fitness.

Working as a team, the medical and safety personnel have to consider the whole person in the total environment—not just on the job.

Programs to rehabilitate alcoholic employees, those with a drug problem, and so on, are becoming more common in plants that at one time simply dismissed an employee who was found to be addicted to alcohol or other drugs.

Some argue that such plants are just going "soft," becoming overindulgent, and inviting trouble. It's true that in the past management took a sterner view; but life was less complicated then, and we understood less about the etiology of these and other problems. Today, every employee is an investment, and it is usually to management's benefit to protect that investment.

First Aid

First aid can be defined as temporary aid and comfort to avert further complications in an emergency until professional medical attention can be obtained. First aid has never been intended to take the place of a physician, nurse, or paramedic. Rather, it should be an integral part of the complete safety program of a plant.

Plants should appoint a safety officer to be responsible for the entire first aid program. This person in turn may enlist supervisory personnel for assistance. It is important, however, to have one individual coordinate the program.

Each department, shift, or work group should have at least one person trained in first aid. The number of trained individuals and the extent of their training will depend on occupational hazards, location of operations in relation to immediate medical attention, and the size of the work unit.

The safety officer should coordinate all first aid training. Free instruction is available from the American Red Cross and some state labor departments. The Red Cross offers several different courses of instruction in first aid, ranging from a very basic course for those who have difficulty with the English language, to advanced courses for first aid instructors. Such trained individuals have proved invaluable to many plant first aid programs.

Training

All employees should be familiar with some first aid techniques such as mouth-to-mouth resuscitation. Supervisors and foremen should have more extensive first aid training.

To assist in meeting the need for personnel trained in first aid as required by the OSHAct regulations, the Secretary of Labor and the president of the American National Red Cross have agreed on the following points:

> Department of Labor recognizes persons who have current Red Cross certificates as fulfilling the regulation.
>
> Red Cross will develop guidelines to expand first aid training to meet the needs of different work environments.
>
> Red Cross will distribute information relating to first aid training services available to employers and employees.

Note. OSHA has entered into an agreement with the American National Red Cross giving OSHA approval to Red Cross first aid training and calling for close cooperative efforts between OSHA and the Red Cross.

There is no requirement that personnel be trained according to Red Cross methods, but only Red Cross training has received specific OSHA approval.

Refresher courses should be taken at least once a year because of changing medical practices, and because company employee rosters are subject to personnel changes that can disrupt the first aid program within a plant or work group. The safety officer should regularly review the list of employees with first aid training certificates.

Educating all employees about the first aid program is essential. They can be kept informed through bulletin boards, newsletters, posters, and so on. Location of first aid kits and the names of employees trained in first aid procedures should be well publicized.

MEDICAL RECORDS AND FACILITIES

The occupational nurse, or the employee in charge of the dispensary, is usually assigned the task of maintaining employees' medical records. These records are kept according to standard procedures, and the job should not be so laborious that it requires most of the time of the person so assigned. Of course the volume may be so great that recordkeeping is the sole responsibility of the person.

Although medical records are confidential, the data should be recorded in standardized language and symbols to be understood by anyone so trained. Standardization also facilitates reliable statistical comparisons and analysis.

To minimize ambiguity and to ensure comparability of terms, records can be made in the terminology of the American Medical Association's *Standard Nomenclature of Diseases and Operations*. The AMA has another useful publication: *Guide to Development of an Industrial Medical Records System*. (For the address of the AMA, see Appendix "Service Organizations and Associations.")

The usual recommendation for a dispensary suggests a minimum of three rooms, a waiting room, a treatment room, and a room for consultation or for making physical examinations. A properly outfitted examining room can serve also for audiometric testing and other purposes. The treatment room should be large enough to treat more than one person at a time.

It is advisable to also have at least one first aid room or station at a location that is convenient to the employees. A room for administering first aid is usually more acceptable, because of the need for privacy.

The first aid room should have an examining table, a cot, a dustproof cabinet for supplies, and a fully equipped first aid kit, a covered waste receptacle, a magnifier with a light, a wheelchair, a stretcher, blankets, a phone with important numbers posted for emergencies, and one or two chairs and a table.

A major step in setting up a first aid program is to select the proper first aid kit. There are many types, designed to fill varying needs, depending on the hazards that might occur. One of the most useful is the unit-type kit. Relatively inexpensive, it is quite suitable for use when nurses are not nearby. The kits have standard-size packages, each containing a single type of dressing or treatment—such as burn cream, ammonia inhalant, eyewash solution, bandage compresses, or aspirin tablets.

No measuring or sorting is necessary, minimizing time and effort when treatment is needed. The unit-type kits are available from several manufacturers and can be furnished to hold 6, 10, 16, 24, or 36 units. Other advantages of the kits are as follows:

> Each unit carton in the kit is overwrapped in a see-through protective covering to keep the contents clean and to indicate whether the carton has been opened.
>
> Each unit carton includes instructions and/or diagrams to help persons use the contents.
>
> All unit-type kit manufacturers offer similar materials packaged in identical size cartons. Restocking from different manufacturers causes no inconvenience.
>
> All contents are immediately visible and easily identified when the kit is opened.
>
> Company logos can be placed on the kits with decals or by silk screening.
>
> Kits are weatherproof and made of steel or high-impact plastic. Each kit has a carrying handle and hangers for wall mounting.

Since OSHA regulations require that all first aid supplies be approved by a consulting industrial physician, it may be helpful to furnish this individual with a complete list of materials available so that he or she can recommend the best combination. Specific requirements are determined by the number of employees and the hazards to which they may be exposed. Once the physician has selected the contents, a certificate of approval should be obtained from him or her and filed for possible future inspection by OSHA compliance officers.

Another key action in instituting a first aid program is determining where to place the first aid kits. They should be located so that no worker in the plant is more than a few minutes away from first aid treatment. Self-evaluation checklists available from the National Safety Council can be helpful in this determination. See self-inspection guide, Table 12-1.

OSHA requires that the color green be used to identify each area designated for first aid, but this rule has been misinterpreted in many ways. Some plants, for example, have painted their first aid kits green. Kits can be any color, but the immediate area in which they are located must be green.

Telephone numbers of selected doctors, hospitals, and ambulances should be placed next to the kits. Such information is particularly helpful to persons working on the night shifts.

Medical Records and Facilities

Table 12-1 Planned Safety and Health Self-Inspection Guide[a]

Department	Unit	Supervisor Responsible	Approved by	Date	Page No.
1. Problems	2. Critical Factors	3. Conditions to Observe	4. Frequency	5. Responsibility	

[a] Principles and Practices of Occupational Safety and Health: A Programmed Instruction Course, OSHA 2215 Student Manual Booklet 3, U.S. Department of Labor, Washington, D.C., p. 39.

A depleted kit is of no use and is a flagrant violation of OSHA. The safety officer has the responsibility to keep each kit well stocked. Unfortunately, a common problem is pilferage. A few special rules, however, can keep pilferage to a minimum:

> All kits should be kept as close as possible to a supervisory area, such as a foreman's desk.
>
> One person, preferably the safety officer, should be responsible for dispensing first aid supplies to individual work areas. The reserve stock should be kept under lock and key at all times.
>
> Employees must be made to realize that unnecessary removal of first aid materials, and even slow replacement of used contents, might mean the difference between life and death.

OSHA regulations prescribe the training of persons in first aid when no infirmary, clinic, or hospital is in proximity to the workplace. Good industrial practice calls for trained first aid personnel in all work situations, even if elaborate facilities are nearby.

The provision of first aid treatment space with at least a cot and supplies is highly desirable, because the work of first aid people is helped by privacy and at least a minimum of equipment.

The regulations leave it entirely to the physician to approve first aid supplies. Various types of kits of supplies are commercially available. The best of supplies require good storage, and a systematic method of replenishing supplies after use or deterioration.

The one type of equipment specifically required by this section of the OSHAct regulations is that designed for quick drenching or flushing of body and eyes. Such facilities must be provided within the work area for immediate emergency use if employees are exposed to injurious corrosive materials.

The regulations do not spell out the details of these facilities, but certain characteristics are well known from experience, especially the experience of the chemical industry.

Eyewash fountains and emergency showers should be so distributed that no exposed worker has to travel far to reach them.

Supplies of water must be adequate (deluge flushing is the rule for emergency showers) and protected against freezing.

Controls should be such that they can be turned on to maximum volume with minimum effort. Persons with acid in their eyes must not be expected to grope for controls, and someone who needs to have corrosives washed off his or her body must not be obliged to understand complicated directions.

Showers and fountains, like fire protection sprinklers, must be protected against ill-considered actions, which might leave them without water supply just when they are needed.

The regulation appears to require fountains and showers in any situation in which exposure to corrosive materials is possible. This includes a host of presently unprotected shops as well as chemical plants, which have long known and used such means of washing off harmful materials.

OSHA MEDICAL TRAINING STANDARDS

The OSHAct regulations for medical facilities are stated in three brief paragraphs. Their brevity should not mislead any employer into considering them insignificant. The full text of the standards promulgated is as follows:

(a) Employer shall insure the ready availability of medical personnel for advice and consultation on matters of plant health.

(b) In the absence of an infirmary, clinic, or hospital in near proximity to the workplace which is used for the treatment of injured employees, a person or persons shall be adequately trained to render first aid. First aid supplies approved by the consulting physician shall be readily available.

(c) Where the eyes or body of any person may be exposed to injurious corrosive materials, suitable facilities for quick drenching or flushing of the eyes and body shall be provided within the work area for immediate emergency use.*

The standard requires that each employer make use of a consulting physician. In a large company or in one whose operations produces acute health hazards, the medical staff might well include one or more physicians, specialists in industrial medicine or in an area related to a hazard peculiar to the industry. In the great majority of companies, the medical consultant will be employed on a part-time retainer basis.

The specific duties assigned to the industrial physician will vary from company to company, but the language of the OSHAct regulation makes it clear that it is not enough to have a physician available only to treat accident victims and employees suffering acute illnesses.

The physician must consult and advise "on matters of plant health" and must approve the first aid supplies provided. The first requirement implies that the doctor must know the plant, its materials, its operations, and the hazards involved. And it is hard to imagine how he or she could meet the requirement of approving first aid supplies without knowing the first aid facilities, qualifications of the first aid personnel, and the types of injury and other health problem that might be expected in the plant.

INDUSTRIAL HYGIENE TRAINING

Medical and industrial hygiene facilities overlap to some extent, and they are certainly interrelated. Hearing conservation is one of the most obvious of the activities that are related to both medicine and hygiene.

Often it is the industrial nurse who heads up the on-the-premises medical team; the physician is available as a consultant and makes examinations, and so on, on a regular schedule.

OSHAct regulations make it imperative that there be interdisciplinary cooperation between the safety professionals, industrial hygienists, and medical personnel.

There are three key concepts in an effective industrial hygiene program—recognition, evaluation, and control.

Recognition requires the knowledge of the stresses arising out of environment, operations, and processes.

*CFR 1910.151.

Evaluation involves a judgment or decision that is usually based on a combination of various measurements of the magnitude of the stress and on past experience.

Control involves some or all of the following: isolating a hazard, substituting a less hazardous process, changing a process, training, housekeeping, the employment of exhaust and ventilation systems, and personal protective equipment.

The stresses creating a hazardous condition fall into one or more of the following categories:

Chemical liquids, gases, dusts, fumes, mists, vapors—that is, respiratory, contact, or ingestion hazards.

Physical—electromagnetic or nonionizing and ionizing radiation, noise, vibration, temperature extremes, and pressure.

Biological—including insects, molds, fungi, and bacteria.

Ergonomic—including matching mental and physical requirements of a job (e.g., operating a large machine such as a lathe) to human capabilities and staying power to minimize monotony, fatigue, and so on.

Of the three modes of entry of a toxic substance—inhalation, ingestion, and absorption—probably the most important is inhalation because of the rapidity with which a toxic material can be absorbed into the bloodstream from the lungs.

A COOPERATIVE TRAINING EFFORT

No single group can solve all the many training problems presented by occupational health hazards. Before any evaluation of an industrial health hazard can be made, there must be an understanding of the whole person. It's not enough to know where and when a hazard occurs on the job and the employee's location or movement in relation to the hazard; there must be some consideration of the employee's off-the-job activities. For example, it's not enough to know how many decibels an employee is exposed to on the job (and for how long each day); it's important to also know what kind of noise the employee is exposed to off the job.

It is difficult to measure off-the-job exposures to hazardous materials or situations, but these exposures cannot be ignored. Some clues come from biochemical measuring and monitoring. But don't overlook the information the employee can relate about him or her self.

Acceptable on-the-job exposures are spelled out for most hazardous substances in the form of threshold limit values (TLVs) published by the

American Conference of Governmental Industrial Hygienists, 1014 Broadway, Cincinnati, Ohio 45202.

Where there is a potential for serious exposure, continuous monitoring is essential. Materials that do not represent too serious a problem, however, may be sampled periodically in accordance with the extent of the problem.

Recognition, evaluation, and control of hazards in an industrial hygiene program depend on cooperation with other disciplines—engineering, medicine, and safety, as well as the purchasing department, line supervision, and the individual employees.

Engineering

The introduction of new plant processes and operations usually begins with the engineering organization. For this reason the engineering staff plays an important role in the control of occupational health hazards. It is part of their responsibility to plan all operations—using established engineering procedures—to prevent unnecessary exposure to harmful work environmental factors or stresses.

Engineers should notify the medical, industrial hygiene, and safety organization whenever new operations or processes are to be initiated. Equally important is a request for an industrial hygiene survey of new installations before permitting employees to operate new equipment or to work with new processes.

Medicine

Medical personnel have the responsibility to know the physical and emotional requirements of every job as well as the hazards involved with each job as factors in the placement of personnel whose physical and emotional capacities meet the minimum job requirements.

The medical team also has an obligation in the development of adequate, effective measures to prevent exposure to harmful agents, to examine all employees periodically—especially those working with or exposed to hazardous materials—and to restrict employees from further exposure (on a medical basis) whenever warranted.

Often the industrial nurse acts as the liaison between the physician and representatives of other disciplines, and occasionally the nurse and the supervisor represent the employees to the other disciplines.

Purchasing

The company's purchasing organization is responsible for making sure that only equipment and materials approved by industrial hygiene, medical, and safety personnel are purchased and that engineering specifications are included in the purchase order.

Supervisors

Each supervisor is responsible for maintaining a safe working environment for employees in his or her charge. In addition to meticulous housekeeping, the supervisor must be aware of the hazards involved in each job, report any accidental exposures to hazardous materials, and know and abide by the work restrictions placed on personnel by the medical department. Supervisors must make sure employees have adequate training, have the proper personal protective equipment, know how to use it, and do, in fact, use it. Discipline and enforcement are usually delegated to the supervisor.

Employees

Each employee is also responsible for contributing to the overall success of a company's industrial hygiene program. Besides following work procedure rules, including the proper use of personal protective equipment, each employee has the obligation to report any accidental exposures or any hazardous conditions that may arise.

Safety

Probably the safety organization plays the key role in the overall environmental health program—that of coordinator.

It is the safety team's basic responsibility to conduct an effective safety program, but this can be done only with the cooperation of engineering, medical, industrial hygiene, purchasing, and supervisory personnel.

Safety's role is educational in that employees must be adequately trained with respect to the hazards involved in their work, regardless of the other requirements of the job. Rules, regulations, procedures, and enforcement are usually initiated by the safety personnel, but much of the responsibility entailed can be delegated to supervisory personnel.

Since new materials and new processes are constantly appearing, an industrial hygiene program remains an ongoing program. Fortunately many organizations as well as consultants specialize in this field. Insurance companies, manufacturers, private laboratories, and consultants can provide a company-wide survey to identify, monitor, and control work environment exposures to health hazards.

13

Evaluating Safety Training Effectiveness

The final chapter presents a broad variety of methods of evaluating overall safety program effectiveness as well as the components of a safety training program. Following a review of some specific measures of safety, selected statistical techniques that have found application in the analysis of safety programs are discussed. Many examples in this important area of safety training are offered.

MEASURES OF SAFETY

Accident Frequency Rate

The accident frequency rate (AFR), also known as the accident rate, is the number of disabling (lost-time) injuries per a specific number of annual worker-hours. For many years the number of worker-hours used was one million and was based on guidance provided by the American National Standards Institute (ANSI) Standard Z16.1. The one million worker-hours standard is based on the hours of work exposure associated with 500 employees working 40 hours each week for 50 weeks each year. More recently OSHA, in recognition that most firms employ far fewer than 500 employees, established a standard based on 100 employees. Today accident frequency rates are calculated using S, the 100 employee standard of 200,000 worker-hours.

$$S = 100 \text{ employees} \times 40 \frac{\text{hours}}{\text{week}} \times 50 \frac{\text{weeks}}{\text{year}}$$

$$= 200,000 \text{ worker-hours}$$

For a firm to determine its own AFR, it need only divide the number of lost-time injuries recorded for a given one-year period (not necessarily a calendar year) by the actual number of worker-hours for that period, and multiply the result by 200,000.

The accident frequency rate is an important value primarily because of its widespread use. Many firms use the AFR to compare the results of their accident history to national averages. These figures are available for all businesses, for all industrial firms, for specific standard industrial codes (SICs), and for various levels of organizational size.

Example 13-1

The Uncommon Medals Company (UMC) has 30 employees at its Oakland facility and 500 employees throughout its organization distributed over 15 locations. In 1978 the Oakland division recorded 5 disabling injuries, and the company overall experienced 60 lost-time injuries. The 1978 national figures for all manufacturing firms in the medals business was 10.0 and varied from a low of 8.0 to a high of 14.0 depending on the size of the firm. Compare UMC to the national results as one way of checking the company's safety training effectiveness.

Solution

$$\text{AFR (for Oakland)} = \frac{5}{(30)(40)(50)} \times 200,000$$

$$= 16.67 \text{ lost-time injuries per OSHA standard for 100 employees}$$

$$\text{AFR (for UMC)} = \frac{60}{(500)(40)(50)} \times 200,000$$

$$= 12.00 \text{ lost-time injuries per OSHA standard for 100 employees}$$

Measures of Safety 187

Comparing Oakland's AFR to UMC's overall AFR, it is readily apparent that Oakland has experienced 38.9% more disabling injuries than the company as a whole.

$$\frac{16.67 - 12.00}{12.00} \times 100 = 38.9\%$$

Furthermore, UMC's AFR is 20.0% higher than the national SIC average for UMC's classification.

$$\frac{12.00 - 10.00}{10.00} \times 100 = 20.0\%$$

Since the statement of the problem does not provide employment levels for the given AFRs, further quantitative analysis cannot be performed.

Review of the given and calculated data indicates that UMC has a potentially undesirable situation if its accident frequency rate statistics are a reliable indicator of the company's safety status. Additionally, there is no question that the Oakland division has experienced a bad year and should receive further examination to ascertain the likelihood of a repeat performance.

More precise judgments that evaluate the company's overall and specific divisional performances can be made using *Accident Facts*, published annually by the National Safety Council.

Accident Severity Rate

The accident severity rate (ASR) is the number of days lost as a result of accidents per a specific number of annual worker-hours. As with the accident frequency rate, the standard for the annual number of worker-hours has dropped from one million to 200,000 since the advent of OSHA. Like the accident frequency rate, the accident severity rate is also important because of its widespread use and it, too, is included in the annual *Accident Facts* booklet.

Example 13-2

The Uncommon Medals Company of Example 13-1 has recorded a total of 125 lost work days for its Oakland division in 1978 and 1200 for the firm overall. The national average for firms with the applicable SIC was 200. Compare the UMC experience both internally and relative to national statistics.

Solution

$$\text{ASR (for Oakland)} = \frac{125}{(30)(40)(50)} \times 200{,}000$$

$$= 416.67 \text{ lost days per OSHA standard for 100 employees}$$

$$\text{ASR (for UMC)} = \frac{1200}{(500)(40)(50)} \times 200{,}000$$

$$= 240.00 \text{ lost days per OSHA standard for 100 employees}$$

Comparing Oakland's ASR to UMC's overall ASR, it is clear that Oakland has experienced 73.6% more lost days than the company as a whole.

$$\frac{416.67 - 240.0}{240.0} \times 100 = 73.6\%$$

Additionally, UMC's ASR is 20.0% higher than the national SIC average for the UMC business classification.

$$\frac{240.0 - 200.0}{200.0} \times 100 = 20.0\%$$

The foregoing quantitative analysis suggests that Oakland has a serious problem. Its accident severity rate is nearly three-quarters again higher than the Uncommon Medals Company overall. This, in combination with its 38.9% higher accident frequency rate, should give UMC management cause for concern. When the company's ASR is coupled with its AFR and compared to national statistics for firms in a similar business, it would be reasonable to conclude that UMC overall is not doing well relative to its SIC and that Oakland, specifically, is a leading indicator of what could happen to the entire company if necessary changes are not made soon.

Frequency—Severity Indicator

Alone, neither the AFR nor the ASR is sufficient to provide an overall indicator of a company's safety performance. There is, however, a way to bring the

Measures of Safety

two together to establish a more valid basis for evaluation. The frequency-severity indicator (FSI) is a single value resulting from appropriately combining the AFR and the ASR. The FSI is calculated using the following equation:

$$FSI = \sqrt{\frac{AFR \times ASR}{1000}} \qquad (13\text{-}1)$$

Example 13-3

Continuing the Uncommon Medals Company example, calculate the FSI for both the Oakland division and for the whole company. Compare the results of these calculations to the FSI values for UMC and Oakland in 1976 and 1977.

Rates	Uncommon Medals	Oakland Division
AFR		
1976	11.50	12.50
1977	13.07	17.02
1978	12.00	16.67
ASR		
1976	200.0	350.0
1977	245.0	405.0
1978	240.0	416.67

Solution

$$FSI\,(1976,\,UMC) = \sqrt{\frac{(11.50)(200.0)}{1000}} = 1.52$$

$$FSI\,(1977,\,UMC) = \sqrt{\frac{(13.07)(245.0)}{1000}} = 1.79$$

$$FSI\,(1978,\,UMC) = \sqrt{\frac{(12.00)(240.0)}{1000}} = 1.70$$

$$FSI\,(1976,\,OD) = \sqrt{\frac{(12.50)(350.0)}{1000}} = 2.09$$

$$\text{FSI (1977, OD)} = \sqrt{\frac{(17.02)(405.0)}{1000}} = 2.63$$

$$\text{FSI (1978, OD)} = \sqrt{\frac{(16.67)(416.67)}{1000}} = 2.64$$

$$\text{FSI (1978, nat'l)} = \sqrt{\frac{(10.0)(200.0)}{1000}} = 1.41$$

Now management has a firmer basis for comparing its safety performance both internally and externally. It is clear that the company overall FSI has been consistently higher than that of comparable firms for the last three years. Furthermore, the Oakland division FSI has been regularly much higher than the company overall. The average FSI for UMC for the three-year period 1976-1978 is 1.67, whereas that for the Oakland division for the same period is 2.45. It can be safely assumed that the 1978 national FSI for the UMC standard industrial code is approximately what a 1976-1978 average would be (i.e., 1.41). Therefore it can be noted that UMC is 18.4% higher than the national average and that the Oakland division is 46.7% higher than UMC.

$$\frac{1.67 - 1.41}{1.41} \times 100 = 18.4\%$$

$$\frac{2.45 - 1.67}{1.67} \times 100 = 46.7\%$$

It can be reasonably concluded that some remedial action is required by UMC management to reverse this undesirable situation.

Average Cost per Injury

The average cost per injury (ACI) combines a number of related factors into a single value, which can then be compared to other ACIs. Costs that are usually considered in developing an ACI value for a business include actual compensation paid to an employee during the calendar period which he or she is incapacitated and cannot work as a result of a job-related injury, medical costs (hospital, doctor, nurse, first aid, ambulance, etc.) paid for by the business in support of a job-related injury, workmen's compensation premiums to the firm's insurance carrier, and others. Comparison of ACIs between departments, from year to year, and so on, provides still another measure of safety.

Example 13-4

The Uncommon Medals Company wishes to examine its average cost per injury as another way to assess the company's safety program. Research of the UMC records reveal the following costs. Calculate the 1978 ACI for their San Diego division (SDD) as well as all three SDD departments, and comment on the differences noted.

1978 Injury Expenses	Dept. A	Dept. B	Dept. C	SDD Overall
Compensation	$10,000	$12,000	$ 6,000	$28,000
Medical costs	7,000	14,000	10,000	31,000
Insurance premiums (pro-rated by number of employees)	5,000	6,000	5,000	16,000
Miscellaneous	2,000	3,000	3,000	8,000
Total	$24,000	$35,000	$24,000	$83,000
1978 injuries	50	75	35	160
1978 ACI	$480	$467	$686	$519

Analysis of the calculations indicates that there is little difference between Departments A and B; however Department C has an ACI 32.2% higher than the overall ACI and 44.9% higher than the average of Departments A and B.

$$\frac{686 - 519}{519} \times 100 = 32.2\%$$

$$\frac{686 - (480 + 467)/2}{(480 + 467)/2} \times 100 = 44.9\%$$

Extending this analysis, management has requested the 1978 ACI be compared to those of previous years. Caution must be exercised when performing year-to-year comparisons that involve dollars. It is necessary to account for inflation to avoid comparison of dollars that are dissimilar in value. Continuing this example, suppose that the following company data are located in the UMC files, as well as associated historical inflation figures gleaned from government sources:

Year	Total Injury-Related Costs	Number of Injuries	Constant $ Value	ACI
1976	$75,000	150	$1.15	$575
1977	$74,000	152	$1.08	$526
1978	$83,000	160	$1.00	$519

where

$$\text{ACI (for 1976)} = \frac{(75{,}000)(1.15)}{150} = \$575$$

$$\text{ACI (for 1977)} = \frac{(74{,}000)(1.08)}{152} = \$526$$

$$\text{ACI (for 1978)} = \frac{(83{,}000)(1.00)}{160} = \$519$$

This analysis indicates that the average cost per injury has decreased 9.74% from 1976 to 1978.

$$\frac{575 - 519}{575} \times 100 = 9.74\%$$

Had management ignored the impact of inflation in its analysis, the ACI for 1976, calculated with the 1976 value dollar, would have been $75,000 ÷ 150 = $500. This inaccurate analysis would have indicated that the ACI had increased by 3.80% from 1976 to 1978.

$$\frac{519 - 500}{500} = 3.80\%$$

Only when the 1978 value of the 1976 dollar is used, as in the preceding tabulation, is an inflation-adjusted figure obtained.

Safe-T-Score

Up to this point, the measures of safety presented have made use of simple, basic math. The Safe-T-Score technique, which employs the results of acci-

Measures of Safety

dent frequency rate calculations, goes beyond basic math. It makes use of a statistical analysis technique referred to as student's t.

$$\text{Safe-T-Score} = \frac{\text{AFR (now)} - \text{AFR (past)}}{\sqrt{\dfrac{\text{AFR (past)}}{\text{worker-hours (now)}/200{,}000}}} \qquad (13\text{-}2)$$

Before demonstrating its use through an example, certain characteristics of Safe-T-Score are worth considering. A Safe-T-Score value does not have any dimensions; that is, it is an artificially created number that is not descriptive of a physical characteristic such as cost (dollars), weight (pounds), or area (square feet). Only the algebraic sign and the value of the Safe-T-Score are of interest to the analyst, a positive sign indicates a worsening situation, and a negative value reflects an improvement of the present situation over the past.

With reference to Figure 13-1, the value of the Safe-T-Score is descriptive of certain conditions as presented in Table 13-1.

Example 13-5

Once again the Uncommon Medals Company is desirous of reviewing its safety performance data to ascertain the status of its safety program. This time UMC has decided to use the Safe-T-Score approach to analyze its inter-

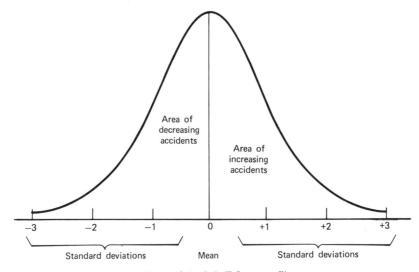

Figure 13-1 Safe-T-Score profile.

Table 13-1 Safe-T-Score Interpretations

Safe-T-Score	Interpretation
Between +3.00 and −3.00 standard deviations	AFR changes are not significantly different. Variation is due to chance.
Over +3.00 standard deviations	AFR is now significantly worse than it was in the past. Investigation of changes are necessary.
Under −3.00 standard deviations	AFR is now significantly better then it was in the past. Considerably fewer accidents are being reported.

nal data. The following 1977–1978 information has been collected for use by an analyst (from Example 13-3).

Data	Oakland Division	UMC (Overall)
1977		
Accidents	4	49
Worker-hours	47,000	750,000
AFR	17.02	13.07
1978		
Accidents	5	60
Worker-hours	60,000	1,000,000
AFR	16.67	12.00

The AFR for the Oakland division has decreased by 2.06% from 17.02 to 16.67, whereas the AFR for the entire company has decreased by 8.19% from 13.07 to 12.00. Without the use of Safe-T-Score it can't be said with certainty whether an event of any statistical significance has occurred. Let's examine the solution.

Solution

Using Eq 13-2, the Safe-T-Score for the Oakland division is calculated as follows:

$$\text{Safe-T-Score (for Oakland)} = \frac{16.67 - 17.02}{\sqrt{\dfrac{17.02}{60,000/200,000}}}$$

$$= \frac{-0.35}{\sqrt{17.02/0.30}} = \frac{-0.35}{\sqrt{56.73}}$$

$$= \frac{-0.35}{7.53} = -0.046$$

Since this Safe-T-Score is very nearly zero, it is determined that virtually no change has taken place.

Again using Equation 13-2, the Safe-T-Score for the entire Uncommon Medals Company is determined:

$$\text{Safe-T-Score (for UMC)} = \frac{12.00 - 13.07}{\sqrt{\dfrac{13.07}{1,000,000/200,000}}}$$

$$= \frac{-1.07}{\sqrt{13.07/5.00}} = \frac{-1.07}{\sqrt{2.61}}$$

$$= \frac{-1.07}{1.62} = -0.66$$

Since this Safe-T-Score is between zero and -3.00, it is reasonable to believe that nothing of any significance has occurred; if something *has* happened, however, we can assume that it is for the better.

Attitude Survey

Earlier in this book it was recommended that an annual employee attitude survey be distributed, collected, and analyzed to assess the effectiveness of a company safety program. Regardless of the questions that are included in the instrument, two characteristics of the survey are valuable tools in the evaluation of a safety program.

Attitude Survey Response Rate

The level of response to an annual survey will vary considerably between companies as well as within a company from year to year. If all other variables are held constant, the survey response rate becomes a thermometer directly measuring the employees' interest and belief in and support of the company safety program. Some of the other variables that must be held as nearly constant as possible include number of questions in the survey, question content, methods of distribution and collection, and time of the year (relative to contract negotiations, payday, holiday, etc.).

Example 13-6

The Uncommon Medals Company annually asks both its hourly and salaried employees to complete an attitude survey instrument that is oriented primarily toward the company safety program. Each year the survey is distributed and collected in a consistent manner, with only minimal changes in question wording to reflect recent changes in the safety program.

The UMC director of safety has been keeping track of the survey results for the five years since it began, as follows:

Year	Number Distributed	Number Returned	Number Usable	Response Rate	Usable Rate
1974	144	72	65	50%	90%
1975	150	84	78	56%	93%
1976	158	104	98	66%	94%
1977	180	108	92	60%	85%
1978	200	146	139	73%	95%

Attempt to relate these results to the company's employee history during this period.

Solution

In 1974 only 50% of the employees returned the survey. In retrospect it appears that this low point represents an attempt by the work force to understand what management was up to with this new feature of the company safety program.

In 1975 and 1976 both the response rate and the usable rate grew dramatically as the employees became more comfortable and more familiar with the instrument.

In 1977, however, these two indicators dropped significantly; a portent of trouble about to surface. Just 60 days after the survey was conducted, union problems nearly precipitated a strike that was narrowly averted. It is possible that UMC management first became aware of the potential problem through a knowledge of the survey results.

In 1978 things appear to have returned to normal with both rates back on their original tracks. Any negative deviation of these trend lines in the future can be used to predict present or developing problems.

Attitude Survey Content Variance

The extent of dispersion surrounding the mean response to a specific question can be a "red flag" indicating considerable variation of opinion regarding the question's content. Minimal variation indicates that there is little difference of opinion about the topic addressed. On the other hand, a high level of dispersion about the mean should be interpreted as an indicator that a consensus does not exist. Whether dispersion is high or low may be difficult to ascertain the first year a survey is conducted. In succeeding years, however, the expanded data base should provide the survey analyst with sufficient background information to make these judgments.

Example 13-7

Continuing the analysis of the Uncommon Medals Company annual employee attitude survey, the following partial results have been documented.

		Year				
Question No.[a]		1974	1975	1976	1977	1978
1.	Mean[b]	3.5	3.6	3.7	3.6	3.7
	S.D.[c]	0.4	0.4	0.3	0.3	0.3
2.	Mean	2.4	2.2	2.4	2.5	2.3
	S.D.	0.5	0.7	0.7	0.8	1.0
3.	Mean	2.1	2.0	2.5	2.7	2.8
	S.D.	0.8	0.6	0.6	0.5	0.4
4.	Mean	4.0	3.7	3.2	3.0	3.0
	S.D.	0.3	0.4	0.6	0.8	0.9
5.	Mean	3.0	3.2	3.5	3.8	4.0
	S.D.	0.5	0.5	0.4	0.3	0.4

[a] The responses to each question are based on a scale of 1 (low) to 5 (high).
[b] The mean (also known as the arithmetic mean, or average), a measure of central tendency.
[c] The standard deviation (S.D.), a statistical measure of dispersion of data around the arithmetic mean.

Review these results in an attempt to identify potential areas of concern.

Solution

Reviewing the data available on a year-by-year basis reveals some important highs and lows in the mean responses to the various questions, but it does not offer much assistance in identifying safety-related personnel problems. A longitudinal analysis, however, offers considerably more insight on several counts.

Question 1 shows high year-to-year consistency in both mean and standard deviation. Question 2, on the other hand, shows a similar level of consistency in its recorded means, but over the five-year period the standard deviation has gradually doubled. This indication of considerable change in the opinions of the respondents serves to identify a potential area for investigation.

Question 3 shows a 33% increase in the mean and a 50% decrease in the standard deviation between 1974 and 1978. Clearly the employees' attitudes on this topic are following an upward trend and, simultaneously, becoming more consistent (less content variance).

Question 4 reflects a situation quite the opposite of Question 3. The mean has dropped by 25% and the variance has increased by a factor of 3. This dramatic change should be cause for concern, and it justifies a thorough investigation of the cause(s).

Finally, the mean of Question 5 reflects a 33% increase (equivalent to Question 3), but the standard deviation is relatively unchanged. This result could be considered to represent an improving trend with little content variance.

The foregoing analysis is a simple example demonstrating the type of information and indicators that are available in reviewing periodic employee attitude surveys beyond the original information expected from the questions.

Hazard Reporting Rate

The hazard reporting rate (HRR) is a function of at least five contributing factors: number of physical hazards that require reporting, number of personnel hazards that require reporting, motivation of personnel to report hazards that have been detected, capacity of personnel to discern the presence of hazards, and the organizational climate (which may or may not be supportive of a high HRR).

The number of physical hazards that require reporting depends primarily on what fraction of a firm's physical hazards have already been reported and the extent of change that occurs within the organization's facilities and equipment. A discussion of physical hazards is found in Chapter 12.

The number of personnel hazards that require reporting is directly related to the extent to which a firm's employees perform their jobs responsibly. When safe job performance is not considered to be paramount, the number of personnel hazards will be excessive.

The level of employees' motivation to report hazards that have been detected is frequently associated with the desirability of the reward system provided by employers and as perceived by the employees. A firm may believe that its reward system is ideally suited to motivate the employees to report hazards; but unless the system is viewed in a similar fashion by the employees, it is unlikely that the employees will be moved to search out and report hazards.

The capacity of personnel to discern the presence of hazards is closely linked to their ability to recognize hazards. This requires considerable attention to training, since most employees are not tuned into identifying hazards that may be in daily proximity to their work areas. A training program especially tailored to upgrading employee hazard recognition skills must be a regular part of a company safety training program.

The organizational climate may or may not be supportive of a high HRR. This factor reflects the extent to which employees in general are cooperating with management to make a firm a safer place to work. When an employee identifies a hazard but neglects to report it despite a desire to receive a reward provided by the employer, it must be assumed that the employee fears condemnation or some other form of mental or even physical punishment by fellow employees. When this is the case, management should be concerned about more than just the reporting of hazards.

Hazard reporting rates can be recorded for analysis in a variety of different formats. Example 13-8 offers some ideas on this subject.

Example 13-8

The director of safety of the Uncommon Medals Company has been keeping records on hazard reporting for three years. Now that a sufficient data base exists, the director has begun to consider some ideas on presentation, to facilitate whatever analysis may eventually be required. Accept the responsibility for this project and offer some tabular methods for data presentation.

Solution

One way to present hazard reporting data is on a gross overall basis as in Table 13-2. This method is easily subdivided into weekly, monthly, or quarterly results, as in Table 13-3.

Another way to look at these figures is by type of hazard, that is, physical or personnel, as in Table 13-4.

Table 13-2 Overall Approach to Hazard Reporting

Year	Number of Hazards Reported
1976	84
1977	96
1978	120

Table 13-3 Periodic Approach to Hazard Reporting

Year	Number of Hazards Reported				
	1st quarter	2nd quarter	3rd quarter	4th quarter	Total
1976	20	20	24	20	84
1977	25	30	35	6	96
1978	30	40	35	15	120

Table 13-4 Categorical Approach to Hazard Reporting

Year	Type of Hazard		Total
	Physical	Personnel	
1976	48	36	48
1977	58	38	96
1978	84	36	120

If the director of safety has reason to be concerned about a specific department's lack of involvement in the hazard reporting system, an arrangement such as that shown in Table 13-5 could be employed.

Naturally, any of these approaches can be combined to offer greater insight on the hazard reporting system. For example, combining the information contained in Tables 13-4 and 13-5 could result in either Table 13-6A or 13-6B, depending on the director's need and/or analysis orientation.

If the departmental sizes are approximately equal, presentations such as the foregoing may suffice to provide usable data for analysis. On the other hand, if the departments are of significantly different sizes, or perhaps the

Table 13-5 Departmental Approach to Hazard Reporting

Year	Department of Reporter			Total
	Dept. X	Dept. Y	Dept. Z	
1976	15	25	44	84
1977	30	20	46	96
1978	45	30	45	120

Table 13-6A Combined Approach to Hazard Reporting: By Department

Year	Department of Reporter								Total
	Dept. X		Dept. Y		Dept. Z		Total		
	Physical Hazard	Personnel Hazard	Physical Hazard	Personnel Hazard	Physical Hazard	Personnel Hazard	Physical Hazard	Personnel Hazard	
1976	10	5	15	10	23	21	48	36	84
1977	20	10	10	10	28	18	58	38	96
1978	25	20	10	20	25	20	84	36	120

Table 13-6B Combined Approach to Hazard Reporting: By Type

Year	Type of Hazard									Total
	Physical Hazard			Personnel Hazard			Total			
	Dept. X	Dept. Y	Dept. Z	Dept. X	Dept. Y	Dept. Z	Dept. X	Dept. Y	Dept. Z	
1976	10	15	23	5	10	21	15	25	44	84
1977	20	10	28	10	10	18	30	20	46	96
1978	25	10	25	20	20	20	45	30	45	120

sizes are changing from one reporting period to another, it is sometimes useful to put the data on a per capita basis. If the departmental sizes were as described in Table 13-7, the information contained in Table 13-5 would appear as in Table 13-8.

The foregoing should not be received as a complete discussion, but rather a beginning approach to hazard reporting analysis.

Table 13-7 Departmental Sizes

	Number of Employees			
Year	Dept. X	Dept. Y	Dept. Z	Total
1976	50	75	125	250
1977	75	100	125	300
1978	100	100	100	300

Table 13-8 Per Capita/Departmental Approach to Hazard Reporting[a]

	Department of Reporter			
Year	Dept. X	Dept. Y	Dept. Z	Total
1976	0.30	0.33	0.35	0.34
1977	0.40	0.20	0.37	0.32
1978	0.45	0.30	0.45	0.40
Average	0.40	0.27	0.39	0.35

[a] Annual reports of hazards per employee.

Facilities Appearance; Housekeeping

Still another measure of effectiveness of a company safety training program is the appearance of the company facilities and equipment; this is frequently referred to as housekeeping.

To be able to evaluate the status of housekeeping within a specific area, a baseline of acceptable attributes must be established (Table 13-9). Once this checklist has been carefully developed, it can serve as a basis for all future evaluations.

Example 13-9

The Uncommon Medals Company has always emphasized the importance of good housekeeping to the company safety program. After many hours of research the director of safety has produced a new, more comprehensive series of housekeeping checklists for use by an persons conducting inspections of specific departments. Each department and specialized work area has its own customized checklist, which is designed to ensure the fairest and most

Measures of Safety

thorough appraisal of housekeeping possible. Produce a customized housekeeping checklist for Department X.

Solution

In Table 13-9, the customized housekeeping checklist for use in Department X, each item should be evaluated and a value assigned for that item for that day. Values can be assigned in any number of ways depending on local preferences.

Table 13-10 offers some ideas for quantitative evaluatory systems for use in conjunction with a housekeeping checklist such as Table 13-9.

Other Measures

Additional measures of safety training effectiveness that may be considered include statistical control charts, workmen's compensation premium ratings, personnel productivity indicators, personnel evaluation ratings, and safety performance indicators. Numerous authoritative references are to be found in the Appendix. These references are logical points to begin a continuing search for other measures that may have application in your organization.

Table 13-9 Customized Housekeeping Checklist

	Department X
	Month October

Item	Date
	1 2 3 4 5 .... 28 29 30 31
1. Portable electrical equipment is grounded	
2. "No Smoking" signs are posted and observed	
3. Fire aisles are kept clear	
.	
.	
.	
9. First aid kit contents are fresh and complete	
10. Hard hats are in use by all personnel in area	
Overall	

Table 13-10 Quantitative Grading Systems for Use with Checklists

Evaluation	Quantitative System					
	I	II	III	IV	V	VI
Excellent	+2	5	4	+3	8	10
Good	+1	4	3	+1	6	8
Average	0	3	2	0	4	6
Fair	−1	2	1	−1	2	4
Poor	−2	1	0	−3	0	2

STATISTICAL TECHNIQUES

Correlation Analysis

Numerous opportunities to use correlation analysis exist in the safety training area. A discussion of this widely used statistical analysis technique can be located in nearly any introductory statistics text.

Some of the areas in which correlation analysis has been successfully used include:

1. Training program examination scores versus personnel evaluation ratings.
2. Training program examination score (average) for a department or shift versus the departmental or shift accident rate.
3. Training program delta scores (the difference between a training program pretest and the final exam) versus personnel evaluation ratings.
4. Training program delta score (average) for a department or shift versus the departmental or shift accident rate.

Example 13-10 demonstrates the use of correlation analysis in the comparison of training program examination scores and personnel evaluation ratings.

Example 13-10

The director of safety of the Uncommon Medals Company is interested in the possible linear relationship between an employee's score on an awareness of

Statistical Techniques

safety knowledge (ASK) examination and that employee's safety rating by his or her supervisor at the completion of the first year of employment. A random sample of seven employees provided the results presented in Table 13-11. Let X be the ASK examination score, and let Y be the supervisor's rating. In each case the maximum possible score (X) or rating (Y) is 15. The director of safety wishes to know whether there is any apparent relationship between X and Y, and if such exists, whether it is linear or nonlinear.

Solution

First, as is appropriate in most statistically oriented problems, we construct a diagram. In this case we use a scatter diagram of the ASK scores on the horizontal (X) axis and of the supervisor's ratings on the vertical (Y) axis (Figure 13-2).

A visual review indicates that a relationship exists; that is, in general as X values increase, Y values also increase. It is not clear from a visual inspection whether a linear relationship exists between the two variables. To determine whether such a relationship actually exists, it is necessary to compute the correlation coefficient r. The mathematical model for r is as follows:

$$r = \frac{n\Sigma XY - \Sigma X \Sigma Y}{\sqrt{n\Sigma X^2 - (\Sigma X)^2} \cdot \sqrt{n\Sigma Y^2 - (\Sigma Y)^2}} \quad (13\text{-}3)$$

where n = the number of data pairs
Σ = sigma (uppercase Greek), which indicates a requirement to sum all values
ΣXY = the sum of all the products resulting from multiplying each XY pairing
ΣX = the sum of all the X values
ΣY = the sum of all the Y values
ΣX^2 = the sum of the squares of all the X values
ΣY^2 = the sum of the squares of all the Y values
$(\Sigma X)^2$ = the squared value of ΣX
$(\Sigma Y)^2$ = the squared value of ΣY

To obtain these values for use in computing r, it is necessary to expand Table 13-11 to show the required calculations. This expansion is presented in Table 13-12.

For Employee No. 1, whose X value was 12 and whose Y value was 11, the product of multiplying 12 times 11 is 132, the XY value. The square of X (i.e., 12) is 144 and of Y (i.e., 11) is 121. This data expansion is carried out

Table 13-11 ASK and Rating Results

Employee	X	Y
1	12	11
2	11	14
3	5	11
4	10	13
5	13	15
6	13	14
7	12	12

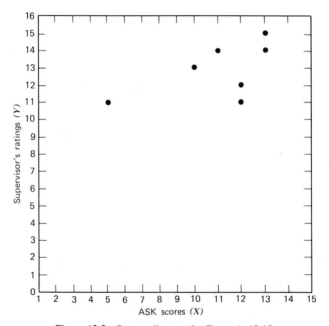

Figure 13-2 Scatter diagram for Example 13-10.

for each employee. When the expansion is complete, each column is summed, to produce the following results:

$$n = 7 \text{ (the number of data pairs)}$$
$$\Sigma XY = 992$$
$$\Sigma X = 76$$
$$\Sigma Y = 90$$

Statistical Techniques

$$\Sigma X^2 = 872$$
$$\Sigma Y^2 = 1172$$
$$(\Sigma X)^2 = (76)^2 = 5776$$
$$(\Sigma Y)^2 = (90)^2 = 8100$$

When applied to the mathematical model for r, these values generate the value of r.

$$r = \frac{7(992) - (76)(90)}{\sqrt{7(872) - 5776} \cdot \sqrt{7(1172) - 8100}}$$

$$= \frac{6944 - 6840}{\sqrt{6104 - 5776} \cdot \sqrt{8204 - 8100}}$$

$$= \frac{104}{\sqrt{328} \cdot \sqrt{104}}$$

$$= \frac{104}{(18.111)(10.198)}$$

$$= \frac{104}{184.694}$$

$$= +0.563$$

This value of r, the correlation coefficient, measures the strength of the linear relationship between X and Y. Since values of r can vary from zero to either

Table 13-12 Example 13-10 Data Expansion

Employee	X	Y	XY	X^2	Y^2
1	12	11	132	144	121
2	11	14	154	121	196
3	5	11	55	25	121
4	10	13	130	100	169
5	13	15	195	169	225
6	13	14	182	169	196
7	12	12	144	144	144
	76	90	992	872	1172

plus or minus one (±1), this analysis indicates a nearly 60% positive correlation in favor of a linear relationship between the ASK test scores and supervisory ratings of employees' safety-related performance. Based on this information, the UMC safety director will very likely decide to continue to use the ASK exam, but will certainly attempt to improve it.

Regression Analysis

Frequent use of regression analysis in the assessment of safety training methods indicates the high regard among safety professionals for the versatility and utility of this approach.

Regression analysis has been successfully applied in numerous safety training-related situations such as the following:

1. Forecasting accident frequency and severity rates for future time periods.
2. Development of mathematical instruments to predict the numerical values of future occurrences.
3. Establishment of linear relationships between previously unrelated variables.
4. Comparison of mathematical relationships between two variables from one time period to another.

Example 13-11 presents regression analysis as a statistical technique being used to develop a mathematical model capable of predicting supervisory ratings based on knowledge of an examination grade.

Example 13-11

The director of safety of the Uncommon Medals Company wishes to use the data collected in Example 13-10 (Table 13-11) in the development of a formula that could be used to predict ratings of employees by their supervisors based on the employees' ASK scores.

The objective is to present a mathematical model or relationship that has the capability to forecast supervisory ratings.

Solution

To determine the desired model, two basic equations are used to determine the slope and the Y intercept of the linear regression line. The equation for the slope m of the line is

$$m = \frac{n\Sigma XY - \Sigma X \Sigma Y}{n\Sigma X^2 - (\Sigma X)^2} \tag{13-4}$$

Statistical Techniques

The slope of a line can be either positive (X and Y both increase or both decrease) or negative (X increases while Y decreases or X decreases while Y increases). The slope is a value that indicates the steepness of the line: that is, a small value of m indicates a line nearly parallel to the X axis and a very large value of m indicates a line nearly parallel to the Y axis. Most lines that are reflective of realistic situations lie somewhere between the two extremes.

Using the findings of Example 13-10, the slope of the desired line is as follows:

$$m = \frac{7(992) - (76)(90)}{7(872) - (76)^2}$$

$$= \frac{6944 - 6840}{6104 - 5776}$$

$$= \frac{104}{328}$$

$$= 0.317$$

The equation for the Y intercept b of the line is

$$b = \frac{\Sigma Y - m\Sigma X}{n} \tag{13-5}$$

The Y intercept of a line is the point on the Y axis where the line crosses it. Depending on whether the line crosses above or below the origin (the intersection of the X and Y axes), it can be either positive (above the origin) or negative (below the origin). It should be noted that in calculating the Y intercept that the value of the slope must be determined first, since m appears in the equation for b.

Using the findings of Example 13-10 and the recently computed value of m, the Y intercept of the desired line is found as follows:

$$b = \frac{90 - 0.317(76)}{7}$$

$$= \frac{90 - 24.092}{7}$$

$$= \frac{65.908}{7}$$

$$= 9.415$$

Therefore the mathematical model for the desired line is

$$Y = 0.317X + 9.415$$

To plot this line it is necessary to know any two points that exist as a part of the line. One is already known—the Y intercept or b.

For this value it is known that when Y is 9.415, X must be zero, since it is at that point that the line intersects or crosses the Y axis. Recall that the Y axis is the dividing line between positive values of X (on the right side) and negative values of X (on the left side). It is at this point that X is always zero. Therefore it is already known that one point on the regression line is (0, 9.415), that is,

$$X = 0$$

$$Y = 9.415$$

The fastest (and easiest) method to calculate a second point on the regression line is to set Y equal to zero in the regression model and algebraically solve the resulting equation for X, as follows:

$$Y = 0.317X + 9.415$$

$$0 = 0.317X + 9.415$$

First, $0.317X$ is subtracted from both sides of the equation to relocate X on the left side of the equals sign.

$$0 - 0.317X = 0.317X - 0.317X + 9.415$$
$$- 0.317X = 9.415$$

Next, both sides of the equation are divided by -0.317 to give X a coefficient of one.

$$\frac{-0.317X}{-0.317} = \frac{9.415}{-0.317}$$

$$X = -29.700$$

Therefore a second point on this regression line is $(-29.700, 0)$, that is,

$$X = -29.700$$
$$Y = 0$$

The line can now be plotted graphically, as in Figure 13-3. First we identify on the regression line the two points that were just calculated. Second, a straightedge placed through the two points specifies the locus of points that comprise the regression line. Figure 13-3 also presents the grouping of original data points on which the regression line was determined.

At this point the UMC director of safety has a new tool to use with the other statistical techniques that can be employed to assess the value of safety training. In addition, forecasts of supervisory ratings are now possible by "plugging in" ASK scores and calculating the likely rating. For example, suppose a new employee has just completed the UMC safety training program. At the conclusion of the program the employee sat for the ASK examination and scored a perfect 15. To determine the supervisory rating this employee is likely to receive, the ASK score is evaluated using the regression relationship.

$$Y = 0.317X + 9.415$$
$$= 0.317(15) + 9.415$$
$$= 4.755 + 9.415$$
$$= 14.17$$

Thus it is concluded that this employee is likely to receive a 14.17 rating by his or her supervisor. When the employee rating is actually rendered, it

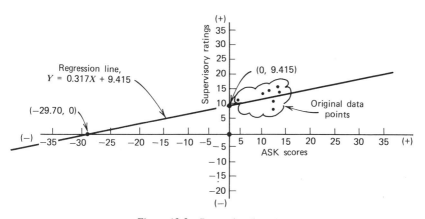

Figure 13-3 Regression line plot.

should be compared to the forecasted rating to continually assess the quality of the training program, the rating procedure, and the regression model.

Chi-Square Analysis

A wide variety of uses of the chi-square (χ^2) analysis technique are found in the safety training field. Some excellent presentations on the derivation and general applications of this nonparametric method can be found in many statistics texts.

A few of the areas in which chi-square analysis has already found successful application in the area of safety training are listed below:

1. Comparison of sample and population frequency distributions (also referred to as a goodness-of-fit test).
2. Comparison of goodness-of-fit for a normal distribution.
3. Test of data independence (also referred to as lack of statistical association).
4. Measures of association in contingency tables.

Examples 13-12 and 13-13 demonstrate two applications of the chi-square technique. The first compares a sample and a known population frequency distribution involving a series of lost-time accidents in a plant; the second addresses measures of association in a contingency table containing accident data related to the amount of safety training received versus supervisory ratings.

Example 13-12

A safety analyst wishes to determine the appropriate frequency distribution to represent the occurrence of lost-time accidents in the Oakland division of the Uncommon Medals Company. The information available to the analyst consists of the numbers of lost-time accidents for a specific period amounting to 100 weeks. During this period 45 weeks had no lost-time accidents, 29 weeks had one lost-time accident, 17 weeks had two lost-time accidents, and 9 weeks had three lost-time accidents. Use a chi-square test to determine the goodness of fit of a Poisson frequency distribution in this situation.

Solution

First, as in any statistical analysis of this type, the null hypothesis (H_0) is established.

$H_0 =$ the observed frequencies are independent and together fit a Poisson distribution

Statistical Techniques

The following observations summarize the given data:

Number of Weeks	Number of Lost-Time Accidents
45	0
29	1
17	2
9	3
100	

Next, the number of degrees of freedom (df) is determined:

$$df = (k - 1) - 1 \tag{13-6}$$

where k is the number of classes, which in this case is 4, and the second "-1" is inserted to account for the estimate of the population mean, μ (the Greek letter mu), by $\overline{X}$, the sample mean.

$$\begin{aligned} df &= (k - 1) - 1 \\ &= (4 - 1) - 1 \\ &= 3 - 1 \\ &= 2 \end{aligned}$$

The 2 degrees of freedom will be used later in the solution to determine the tabulated value of χ^2 (chi square).

Now the estimate ($\overline{X}$) of the population mean is calculated by multiplying each of the observed number of weeks by its corresponding observed number of lost-time accidents and summing the four products. This sum is then divided by the total number of weeks.

$$\begin{aligned} \overline{X} &= \frac{(45)(0) + (29)(1) + (17)(2) + (9)(3)}{100} \\ &= \frac{0 + 29 + 34 + 27}{100} \\ &= \frac{90}{100} \\ \overline{X} &= 0.90 \end{aligned}$$

This becomes our estimate of μ.

Next we enter a Poisson distribution table (found in the back of many introductory statistics texts) for $\mu \approx \overline{X} = 0.90$ and extract the following probabilities.

A	P(A)
0	0.4066
1	0.3659
2	0.1647
3	0.0494
4	0.0111
5	0.0020
6	0.0003
	1.0000

where A symbolizes the number of lost-time accidents and $P(A)$ is the probability of the occurrence of that number.

Since data for 100 weeks are under analysis, multiplying 100 times each accident level probability produces the theoretical or expected number of weeks associated with each accident level.

A	P(A)	Expected Number of Weeks[a]
0	0.4066	40.66
1	0.3659	36.59
2	0.1647	16.47
3	0.0494	4.94 ⎫
4	0.0111	1.11 ⎬ 6.28
5	0.0020	0.20 ⎪
6	0.0003	0.03 ⎭
	1.0000	100.00

[a] Expected number of weeks = $100 \cdot P(A)$.

Statistical theory requires that no value in the expected column be less than 5. When a value less than 5 is identified, it is appropriate to combine it with other expected values so as to meet this constraint. In this case the last four classes are combined to total 6.28.

Statistical Techniques

In preparation for the chi-square analysis, a complete set of expected and observed values is assembled. This facilitates the development of the calculated chi-square value.

A	P(A)	E, Expected Number of Weeks	O, Observed Number of Weeks
0	0.4066	40.66	45
1	0.3659	36.59	29
2	0.1647	16.47	17
3	0.0494	4.94 ⎫	9 ⎫
4	0.0111	1.11 ⎬ 6.28	0 ⎬ 9
5	0.0020	0.20 ⎪	0 ⎪
6	0.0003	0.03 ⎭	0 ⎭
	1.0000	100.00	100

To facilitate the chi-square calculation, the last four classes of the observed number of weeks will also be combined. This leaves four classes of accident levels to be evaluated: 0, 1, 2, and 3 to 6.

The mathematical model for determining the calculated (calc) value of chi-square is as follows:

$$\chi^2_{calc} = \Sigma \frac{(O - E)^2}{E} \qquad (13\text{-}7)$$

$$\chi^2_{calc} = \frac{(45 - 40.66)^2}{40.66} + \frac{(29 - 36.59)^2}{36.59} + \frac{(17 - 16.47)^2}{16.47} + \frac{(9 - 6.28)^2}{6.28}$$

$$= 0.4632 + 1.5744 + 0.0171 + 1.1781$$

$$\chi^2_{calc} = 3.2328$$

Assuming that a confidence level (CL) of 95% is sufficient to meet the analyst's needs, and that the 2 degrees of freedom calculated earlier is appropriate, a tabulated (tab) value of chi-square is determined. In statistics texts the chi-square table is usually found near the Poisson frequency table mentioned earlier in this example.

$$\chi^2_{tab} = \chi^2_{\substack{df=2 \\ CL=95\%}} = 5.9915$$

Finally, χ^2_{calc} is compared to χ^2_{tab}. In this case χ^2_{calc} is less than χ^2_{tab}; therefore H_0 is accepted and the safety analyst has a high degree (95%) of assurance that the frequency distribution does in fact follow the Poisson model. For this reason the Poisson frequency distribution can be used in the future analysis of lost-time accidents in the Uncommon Medals Oakland division. This knowledge will facilitate the analyst's understanding of the existing accident process and will provide a basis for comparison, should the system change.

Example 13-13

In a study of the possible relationship between the amount of safety training given to production line workers and their safety-based ratings (rendered by their supervisors) after working for six months, the following results were obtained:

		Ratings		
		Excellent	Good	Poor
Amount of Safety Training	1 day	6	12	0
	3 days	12	25	6
	5 days	14	31	12
	10 days	2	23	7

The safety analyst of the Uncommon Medals Company desires to determine whether there is significant association at the 0.01 level (confidence level is 99%) between the amount of safety training and the supervisors' ratings, according to this information.

Solution

The null hypothesis (H_0) for this example is expressed as follows:

H_0 = worker ratings are independent of the amount of safety training they receive (i.e., there is no relationship between ratings and training)

The analysis is begun by developing horizontal and vertical totals for the given matrix of observed values.

Statistical Techniques

	Excellent	Good	Poor	Total
1 day	6	12	0	18
3 days	12	25	6	43
5 days	14	31	12	57
10 days	2	23	7	32
	34	91	25	150

Next the expected value will be determined for each row-column intersection. For example for the "1 day—Excellent" intersection the calculation is

$$\frac{\text{(total of the ``1 day'' row) (total of the ``excellent'' column)}}{\text{total of all rows and columns}}$$

$$\frac{(18)(34)}{150} = 4.08$$

Therefore 4.08 is the expected value, which corresponds to the observed value of 6.

One more example; the expected value for the "5 days—Good" intersection is

$$\frac{(57)(91)}{150} = 34.58$$

Therefore 34.58 is the expected value, which corresponds to the observed value of 31.

Completing this series of calculations and inserting the expected values to the right of the observed values produces the following matrix.

	Excellent	Good	Poor	Total
1 day	6/4.08	12/10.92	0/3.00	18/18
2 days	12/9.74	25/26.09	6/7.17	43/43
5 days	14/12.92	31/34.58	12/9.50	57/57
10 days	2/7.26	23/19.41	7/5.33	32/32
	34/34	91/91	25/25	150/150

Calculation of the chi-square value requires use of the standard mathematical model presented in Example 13-11.

$$\chi_{calc}^2 = \Sigma \frac{(O - E)^2}{E} \qquad (13\text{-}7)$$

In this case 12 individual factors will be determined and summed to produce χ^2_{calc}.

$$\chi^2_{calc} = \frac{(6 - 4.08)^2}{4.08} + \frac{(12 - 10.92)^2}{10.92} + \frac{(0 - 3.00)^2}{3.00}$$

$$+ \frac{(12 - 9.74)^2}{9.74} + \frac{(25 - 26.09)^2}{26.09}$$

$$+ \frac{(6 - 7.17)^2}{7.17} + \frac{(14 - 12.92)^2}{12.92} + \frac{(31 - 34.58)^2}{34.58}$$

$$+ \frac{(12 - 9.50)^2}{9.50} + \frac{(2 - 7.26)^2}{7.26}$$

$$+ \frac{(23 - 19.41)^2}{19.41} + \frac{(7 - 5.33)^2}{5.33}$$

$$= 0.9035 + 0.1068 + 3.0000 + 0.5244 + 0.0455$$
$$+ 0.1909 + 0.9028 + 0.3706 + 0.6597 + 3.8110$$
$$+ 0.6640 + 0.5232$$

$$= 11.7006$$

To determine the tabulated value of χ^2, the number of degrees of freedom is needed.

$$df = (\text{number of rows} - 1)(\text{number of columns} - 1) \qquad (13\text{-}8)$$
$$= (4 - 1)(3 - 1)$$
$$= (3)(2)$$
$$= 6$$

Based on the requirement to use a significance level of 0.01 (confidence level = 99%) and $df = 6$, a tabulated value of χ^2 is determined. Chi-square tables are frequently found in introductory statistics texts.

Statistical Techniques

$$\chi^2{}_{tab} = \chi^2{}_{\substack{df=6 \\ CL=99\%}} = 16.8119$$

Since $\chi^2{}_{calc}$ is less than $\chi^2{}_{tab}$, H_0 cannot be rejected. Therefore it can be concluded at the 99% confidence level that there is no significant association between the amount of safety training and the new employees' ratings, according to the given data.

This disappointing news should serve as an incentive to the UMC director of safety to review both the content of the safety training program (and the instructors) and the form and content of the supervisors' employee rating form. A good training program and a good form would have indicated a high degree of statistical association.

Probabilistic Analysis

One of the most versatile statistical methods available in measuring the effectiveness of safety training is probabilistic analysis. Virtually every statistics text has extensive coverage devoted to this most basic, yet highly important technique.

Some areas in which probabilistic analysis has been successfully applied in measuring the effectiveness of safety training methods include the following:

1. Determination of likelihood of increasing safe performance by altering safety training program.
2. Assessment of accident probabilities associated with a variety of machine guarding configurations.
3. Development of probabilities related to discovery of occupational diseases given the utilization of selected medical procedures.
4. Calculation of forecasted likelihoods of events occurring given the probabilities of selected prior events.

Example 13-14 demonstrates how probabilistic analysis can be used to assess the impact of alterations to an existing safety program.

Example 13-14

The probability that a specific change (CA) to the existing Uncommon Medals Company safety training program will increase the number of accident-free days by 20% is 0.40, and the probability and it will have no effect on the number of accident-free days is 0.60. Based on the experience of other firms in the medals industry, there is no chance that its implementation will be the cause of fewer accident-free days.

Also, the probability that another change (*CB*) will increase the number of accident-free days by the same 20% is 0.30, and the probability that it will have no effect is 0.70. There is no likelihood of a negative impact.

Finally, the probability of a third change (*CC*) to the existing UMC safety training program increasing the number of accident-free days by 10% is 0.60; that there is a 0.40 probability of no effect.

Previous experience using the same changes in similar companies indicates that their effects on the number of accident-free days are statistically independent. That is, their effects are cumulative, not overlapping in the event that more than one change is implemented.

If all three procedures are implemented simultaneously, the UMC director of safety needs to know the probabilities that each of the following will occur:

A. There will be a 30% increase in the number of accident-free days.
B. There will be a 40% increase in the number of accident-free days.
C. There will be *no* increase in the number of accident-free days.

These probabilities are required to support further calculations related to UMC cost-benefit analyses.

Solution

As with nearly any mathematically oriented problem, it is very useful to rewrite the problem statement in quantitative terms.

Statement Number	Statement
A1	$P(20\% \text{ increase}/CA) = 0.40$
A2	$P(0\% \text{ increase}/CA) = 0.60$
B1	$P(20\% \text{ increase}/CB) = 0.30$
B2	$P(0\% \text{ increase}/CB) = 0.70$
C1	$P(10\% \text{ increase}/CC) = 0.60$
C2	$P(0\% \text{ increase}/CC) = 0.40$

Statement A1 should be read as follows: the probability of a 20% increase in the number of accident-free days, given that change *A* has been implemented, is 0.40.

Statement A2 should be read as follows: the probability of a 0% increase in the number of accident-free days, given that change *A* has been implemented, is 0.60.

Statistical Techniques   221

The remaining statements (B1, B2, C1 and C2) are all read in a similar way.

A. In response to the UMC safety director's first concern, the analysis is as follows:

The probability of a 30% increase in the number of accident-free days when all three changes are implemented simultaneously is the sum of two conditions: when change A causes a 20% increase, change B causes a 0% increase and change C causes a 10% increase; and when change A causes a 0% increase, change B a 20% increase, and change C a 10% increase.

This combination of conditions and events is computed as follows:

$P(30\% \text{ increase}/CA, CB, CC)$
$= A1 \cdot B2 \cdot C1 + A2 \cdot B1 \cdot C1$
$= (0.4)(0.7)(0.6) + (0.6)(0.3)(0.6)$
$= 0.168 + 0.108$
$= 0.276$

Therefore there is nearly a 28% chance of a 30% increase in the number of accident-free days if all three changes are implemented simultaneously.

B. With respect to the second area of interest, the analysis is conducted, as follows:

The probability of a 40% increase in the number of accident-free days when all three changes are brought on-line together is the product of three events: when change A causes a 20% increase, when change B causes a 20% increase, and when change C causes no increase at all.

This combination of events is computed as follows:

$P(40\% \text{ increase}/CA, CB, CC)$
$= A1 \cdot B1 \cdot C2$
$= (0.4)(0.3)(0.4)$
$= 0.048$

Therefore there is just less than a 5% chance of a 40% increase in the number of accident-free days if all three changes are implemented simultaneously.

C. Finally, addressing the safety director's third concern, we have the following analysis:

The probability of a 0% increase in the number of accident-free days associated with the simultaneous incorporation of all three changes in the existing UMC safety training program is also the product of three events: when change A causes a 0% increase, when change B causes a 0% increase, and change C causes a 0% increase.

This product is calculated as follows:

$P(0\% \text{ increase}/CA, CB, CC)$
$= A2 \cdot B2 \cdot C2$
$= (0.6)(0.7)(0.4)$
$= 0.168$

Therefore there is nearly a 17% chance of absolutely no increase in the number of accident-free days if all three changes are brought into use together.

Had the UMC safety director been interested in the probability of a 50% increase in the number of accident-free days associated with the simultaneous implementation of changes A, B, and C, analysis would have revealed a 7.2% chance of such an increase.

Table 13-13 summarizes the findings associated with this example and Figure 13-4 portrays these results graphically. Several additional conditions

Table 13-13 Summary of Findings

X, Increase in Number of Accident-Free Days	Y, Likelihood of Occurrence
0%	16.8%
10%	25.2%
20%	18.4%
30%	27.6%
40%	4.8%
50%	7.2%

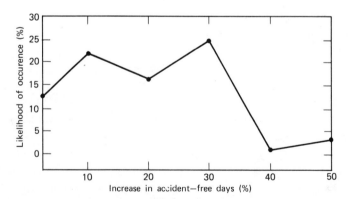

Figure 13-4 Presentation of findings.

are included to complete the full scope of possibilities. The careful reader will note that the likelihoods of occurrence sum to 100%, since the conditions are mutually exclusive (if one occurs, it precludes the possibility of any other), collectively exhaustive (no condition can exist other than those considered), and independent of each other.

These results are quite important in that the likelihoods will be applied in subsequent cost-benefit calculations to justify or withdraw support for any or all of the three changes to the safety training program.

Other Statistical Techniques

Many statistical techniques are available for use in measuring the effectiveness of a safety training program in addition to correlation analysis, regression analysis, chi-square analysis, and probabilistic analysis.

Other statistical techniques that are already in regular use in the analysis of safety training methods include the following:

Analysis of Variance. This is a simple method for the simultaneous comparison of several treatments to make inferences about the relationship between the treatments and the particular variable to be measured.

Safety Sampling. This technique is based on the quality control approach to analysis of output, that is, random inspection rather than 100% checking of every item.

SCRAPE. This is a simple procedure designed to provide management with periodic informational up-dates on a company's accident prevention performance.

Job Site Survey. This is a regular managerial analysis of ongoing jobs to assess the overall quality of the safety program in each area of concern.

Appendix

DEFINITIONS*

abate	To correct under the Occupational Safety and Health Act; to come into compliance with a standard that is being violated.
citation	Issued by the OSHA Area Director to an employer for an alleged violation reported by the OSHA Compliance and Safety and Health Officer during a compliance visit.
comply	To act in accordance with the Occupational Safety and Health Standards; to follow the rules and regulations published in the Code of Federal Regulations.
contest	To object to or to appeal a decision made by the OSHA Area Director.
de minimis violation	Violation of a standard that does not involve an immediate or direct relationship to the safety or health of an employee.
hazard	A risk, danger, or peril to employees in the workplace.
imminent danger	Any condition or practice in any place of employment which is such that a danger exists which could reasonably be expected to cause death or serious physical harm immediately or before the imminence of such danger can be eliminated through the enforcement procedures otherwise provided by the Act.

---

*Principles and Practices of Occupational Safety and Health: A Programmed Instruction Course. OSHA 2213, Student Manual Booklet 1, U.S. Department of Labor, Washington, D.C., p. 29.

Model Safety Program 225

promulgate To issue, establish, or make known officially the terms of a law or regulation having the force of law. For example, to publish OSHA rules, procedures, standards, and regulations in the *Federal Register*.

"recognized" hazard A hazard is recognized if it is a condition that is (*a*) of common knowledge or general recognition in the particular industry in which it occurs, and (*b*) detectable (1) by means of senses (sight, smell, touch, and hearing), or (2) is of such wide, general recognition as a hazard in the industry that even if it is not detectable by means of the senses, there are generally known and accepted tests for its existence which should make its presence known to the employer. For example, excessive concentrations of a toxic substance in the air would be a "recognized" hazard even though they could be detected only through the use of measuring devices.

repeated violation A violation of any standard, rule or order, or the general duty clause, may be cited as repeated under the Act (Section 17a) where a second citation is issued under the Act for violation of the same standard, rule, or order, or the same condition violating the general duty clause for which a previous citation was issued. A repeated violation differs from a failure to abate in that repeated violations exist where the employer has abated an earlier violation and, upon later inspection, is found to have violated the same standard. A notice of failure to abate would be appropriate where the employer has been cited and fails to abate the hazard cited within the abatement period.

standard A rule, established in accordance with law or other competent authority, which designates safe and healthful conditions or practices by which work must be performed to prevent injury or illness.

MODEL SAFETY PROGRAM

The XYZ Company director of safety is in complete charge of the company safety program. The duties of the director of safety include a number of additional responsibilities. These are listed below.

1. Appoint a chairperson to direct the activities of the employee safety committee.
2. See that each major department in the company has representation on the safety committee.
    a. Each member of the committee should be selected by the personnel he or she represents and should be seated for no less than one year and no more than two years.
    b. Seat assignments should be staggered according to time so that no more than one-third of the committee will be new members at any time.
    c. Total membership on the committee should not exceed nine employees.
    d. The committee chairperson will be responsible for providing meeting agendas three days in advance of each meeting and minutes of each meeting within five days following a meeting.
    e. The committee chairperson will provide a copy of the minutes of each monthly meeting to the director of safety and the plant manager.
    f. The safety committee should be responsible for the following:
        (1) Keeping fellow employees aware of what the safety committee is doing.
        (2) Determining solutions to new safety and health problems as they arise.
        (3) Keeping abreast of the latest changes in personnel protective equipment, general safety information, and laws pertaining to employee safety and health.
        (4) Reviewing employee charges of failure to abate safety and health hazards by maintenance personnel.
        (5) Reviewing all accidents in detail to determine the responsible conditions.
        (6) Developing whatever corrective actions are necessary to eliminate conditions that have been determined as being responsible for an accident.
        (7) Following up to verify that corrective actions to eliminate conditions that have caused an accident have been completed.
3. Direct the director of maintenance (or plant engineer) to make frequent, regular safety inspections of the entire company facility and to record all findings. A copy of each such report should be retained by the director of safety.

Model Safety Program

- a. The director of maintenance (or plant engineer) should be held responsible for staying completely aware of all the safety hazards and should know about OSHA violations that exist within the company facility.
- b. The director of maintenance (or plant engineer) should be totally knowledgeable of the pertinent requirements of the latest edition of the *Federal Register* containing OSHA specifications.
- c. The director of maintenance (or plant engineer) should be provided adequate opportunity by the company to attend safety workshops or seminars where he or she can become aware of all pertinent OSHA standards. The week-long course offered by OSHA in the Rosemont, Illinois training facility is an example.

4. Direct the director of maintenance (or plant engineer) to make repairs or maintenance related to safety or health hazard abatements a number one priority.
   - a. A regular system whereby any employee can submit a notification of a safety or health hazard should be established. All notices should be submitted through the office of the director of safety (where a record of the notice is made) to the director of maintenance (or plant engineer). A company form entitled "Request for Corrective Action" could be used for this purpose.
   - b. Written explanation of failure to complete such abatement within two working days should be required.

   This improved project identification and tracking procedure should ensure that specific safety-related projects are located, responsibility for abatement assigned, and follow-up to determine quality and extent of project completion is conducted.

5. Safety literature such as brochures, movies, posters, and notices, should be obtained and disseminated as widely as possible on a regular basis.
   - a. Safety bulletin boards should be established throughout the company facility, and material posted thereon should be kept clean and current.
   - b. Speakers who are specialists in selected areas of safety and health should be invited in to meet with employees of the company.

6. In the area of records maintenance, the director of safety should be responsible for keeping current files on the following:
   - a. **Accidents.** OSHA Forms 200 and 101 must be used consistent with the OSHA recordkeeping requirements pamphlet.
   - b. **Inspections.** Records of all regular and spot inspections performed by the director of maintenance (or plant engineer) or his or

her designated representative should be kept in a separate volume with the equipment check records.

 c. **Equipment Checks.** Reviews of fire extinguishers, hot water heaters, air compressors, air conditioners, and heating units, should be performed by the director of maintenance (or plant engineer) or his or her designated representative and records of these checks maintained along with the inspection records.

 d. Copies of all safety meeting agendas and minutes should be maintained in a separate volume. Minutes should contain the following information:
  (1) Dates.
  (2) Times.
  (3) Attendance (by name, department, and shift).
  (4) Topics discussed.
  (5) Resolutions.
  (6) Special features (e.g., speakers, films, etc.).

 e. Records of all internal (company) and external correspondence related to employee safety and health, including the following:
  (1) Requisitions for equipment with the statement that such equipment meets OSHA standards.
  (2) Letters requesting films, brochures, posters, and so on.
  (3) Notices to employees concerning company policy changes, usage of protective equipment, temporary hazards during modification or construction inside or outside of the company facilities, and so on.

7. Maintain as many complete physician-approved first aid kits as are needed and a list of employees with current (within three years of certification) Red Cross first aid training.
 a. Schedule employees for recertification as necessary.
 b. Restock the first aid kits no less than monthly to ensure a complete supply of required materials.

Orientation Program

Orientation or introductory training is primarily concerned with acqainting newly employed workers with their surroundings. It should being with an official welcome or greeting from the director of personnel followed by a few minutes of informal chatting. If there are several new employees to be introduced to the company on any one day, the personnel director ordinarily greets them as a group.

Model Safety Program

Although this meeting with new employees is often described as an "induction interview," there is little interviewing done. On the contrary, the director of personnel does most of the talking in attempting to acquaint the workers with their new surroundings. These talks are friendly, but brief.

The topics generally follow a sequence similar to the listing below, which some personnel directors prefer to jot down, to be sure that none of the items is overlooked.

1. Put employee(s) at ease.
    a. Greet them pleasantly.
    b. Ask about interests and experience.
2. Discuss wages and hours.
    a. Rate of pay—standard and overtime.
    b. Place, time, and method of pay.
    c. How and where to "punch in" or record working time.
    d. Times—starting, quitting, lunch, and breaks.
    e. Opportunity for promotion and pay raises.
3. Location of company facilities.
    a. Washrooms.
    b. Locker rooms.
    c. Toilets.
    d. Exits.
    e. First aid dispensary.
    f. Lunch room.
    g. Personnel department.
    h. Fire extinguishers.
4. Company policies.
    a. Reporting absences and late arrivals.
    b. Grounds for discharge or discipline.
    c. Safety requirements.
5. Safety regulations.
    a. Provide company safety manual and review contents.*
    b. Job safety rules.

---

*The director of personnel is responsible to the director of safety for the establishment and maintenance of documentation to assure that all new employees received the company safety manual and that they had an opportunity to review any portion of the manual.

c. Safety committee.
   d. Prompt reports of injuries.
   e. Use of personnel protection guards.
   f. Need for good housekeeping.
6. Company benefits (important and uncommon ones).
7. Company plans for future (to indicate progressive nature of company).
8. Trip around plant.
9. Introduce each new employee to his or her immediate supervisor and fellow employees.
10. Job instruction by the immediate supervisor (provide copy of job description).

With reference to the final item on the checklist, the director of personnel should have previously confirmed the ability of each first-line supervisor to clearly communicate the job instruction phase of the orientation program. This phase can be expressed in four steps.

1. *Tell* the employee how to do the job.
2. *Show* the employee how to do the job.
3. Have the employee *try* to do the job to ensure complete understanding.
4. *Check* the employee through close supervision to be sure that he or she continues to follow instructions after being authorized to *begin* an assignment.

Safety Incentives Program

The development and utilization of a safety incentives program to reward safe, accident-free behavior by cooperating employees is an important part of the XYZ Company safety program. The company offers its employees of over one year an opportunity to share in the reduced costs of workmen's compensation insurance premiums that are associated with reduced accident rates and reduction in the number of lost days related to job-connected accidents.

The company uses the following chart for determining the value for a cash award for accident-free job behavior (this material is an example only and is not specifically recommended for adoption):

Model Safety Program

Time	Award
1-5 years	$20 each year plus $100 additional for five consecutive years
6-10 years	$25 each year plus $150 additional for five more consecutive years
11-15 years	$30 each year plus $200 additional for five more consecutive years
16-20 years	$35 each year plus $250 additional for five more consecutive years
21-25 years	$40 each year plus $300 additional for five consecutive years
Over 26 years	$50 each year

The initial counting period begins with each employee's first day of work with the company. Thereafter, should an employee be found to be the cause of an accident to himself or herself or to be involved in an accident that is determined to be the fault of the employee, the next counting period should begin with the next regular working day following the accident.

The cash awards will all be made during the first week of December of each year so that the money will become available at a convenient time. The award will be in the form of a special check, it is not to be included as a part of the regular paycheck. The award presentations will be made during a scheduled employee gathering, either solely for this purpose or in conjunction with other safety activities.

Finally it is recommended that when the OSHA Form 200 is posted in the early part of each year, a notice of employees who have received accident-free awards should be posted adjacent to it. The company will identify each individual by his or her organizational unit, for example:

XYZ Company
Safety Awards Roster
1979

Maintenance	
Joe Blow	3 Years
John Smith	7 Years
Quality Control	
Bill Jones	12 Years
Packaging	
Sam Brown	1 Year

SERVICE ORGANIZATIONS AND ASSOCIATIONS

The following list will provide you with an idea of the types of services available to the general public as well as to members of these organizations. A more complete list of possible sources can be found in the National Safety Council publication titled "Accident Prevention Manual for Industrial Operations."

American Chemical Society
1155 16th Street, N.W.
Washington, D.C. 20036

- This society has a committee on chemical safety

American Industrial Hygiene Association
210 Haddon Avenue
Westmont, New Jersey 08108

- This association will furnish names of industrial hygienists in your area

American Medical Association
Department of Occupational Health
535 North Dearborn Street
Chicago, Illinois 60610

- This association has many pamphlets on occupational health subjects

American National Standards Institute
1430 Broadway
New York, New York 10018

- Many standards set by this organization were adopted as OSHA's initial standards

American National Red Cross
Safety Services
17th and D Streets, N.W.
Washington, D.C. 20006

- This organization has developed training programs that will help your establishment meet the first aid requirements listed in the standards

American Public Health Association
1740 Broadway
New York, New York 10019

- A committee of this association deals with injury control and emergency services

American Society for Testing and Materials
1916 Race Street
Philadelphia, Pennsylvania 19103

- The society sponsors research in the properties of engineering materials and develops standards, including specifications and test methods

American Society of Safety Engineers
850 Busse Highway
Park Ridge, Illinois 60068

- The society promotes and develops educational programs for safety training and conducts research in safety areas

Human Factors Society
P.O. Box 1369
Santa Monica, California 90406

- This society will help in the referral of human factors specialists on request

Industrial Hygiene Foundation of America, Inc.
5231 Centre Avenue
Pittsburgh, Pennsylvania 15232

- Will assist establishments in the development of health programs

Industrial Medical Association
55 East Washington Street
Chicago, Illinois 60602

- This association sponsors committees in areas such as industrial hygiene and clinical toxicology, radiation, and education and training

Industrial Safety Equipment Association, Inc.
60 East 42nd Street
New York, New York 10017

- Will provide information on personal protective equipment for industry

The National Fire Protection Association
60 Batterymarch Street
Boston, Massachusetts 02110

- A clearinghouse on the subjects of fire prevention and protection

The National Safety Council
425 North Michigan Avenue
Chicago, Illinois 60611

- The largest organization in the world devoted to the prevention of injury; accident prevention material and programs are available through this council

National Society for the Prevention of Blindness, Inc.
79 Madison Avenue
New York, New York 10016

- Participates as a member of the American National Standards Institute in studies on illumination, vision, and eye protection

Underwriters Laboratories, Inc.
207 East Ohio Street
Chicago, Illinois 60611

- Maintain laboratories for the examination and testing of devices, materials and systems

U.S. DEPARTMENT OF LABOR, OFFICE OF THE SOLICITOR: REGIONAL OFFICES

Region 1—Boston

Regional Solicitor
U.S. Department of Labor
John F. Kennedy Federal Building
Government Center, Room 1607
Boston, Massachusetts 02203

Region 2—New York

Regional Solicitor
U.S. Department of Labor
Parcel Post Building
341 Ninth Avenue, Room 900
New York, New York 10001

Regional Attorney
U.S. Department of Labor
Box 13344
Santurce, Puerto Rico 00908

Region 3—Philadelphia

Regional Solicitor
U.S. Department of Labor
Jefferson Building
1015 Chestnut Street
Philadelphia, Pennsylvania 19107

Region 4—Atlanta

Regional Solicitor
U.S. Department of Labor
1371 Peachtree Street, N.E., Room 339
Atlanta, Georgia, 30309

Associate Regional Solicitor
U.S. Department of Labor
1929 Ninth Avenue, South
Birmingham, Alabama 53205

Regional Attorney
U.S. Department of Labor
U.S. Court House Building
801 Broad Street, Room 725
Nashville, Tennessee 37203

Region 5—Chicago

Regional Solicitor
U.S. Department of Labor
Everett McKinley Dirksen Building
219 South Dearborn Street—Room 712
Chicago, Illinois 60604

Regional Attorney
U.S. Department of Labor
Federal Office Building, Room 881
1240 East Ninth Street
Cleveland, Ohio 44199

Associate Regional Attorney
U.S. Department of Labor
1502 Washington Blvd. Building
234 State Street
Detroit, Michigan 48226

Region 6—Dallas

Regional Solicitor
U.S. Department of Labor
Federal Building and U.S. Court House, Room 7052
1100 Commerce Street
Dallas, Texas 75202

Region 7—Kansas City

Regional Solicitor
U.S. Department of Labor
2106 Federal Office Building
911 Walnut Street
Kansas City, Missouri 64106

Region 8—Denver

Attorney-in-Charge
U.S. Department of Labor
Federal Office Building, Room 16444
1961 Stout Street
Denver, Colorado 80202

Region 9—San Francisco

Regional Solicitor
U.S. Department of Labor
Federal Building, P.O. Box 36017
450 Golden Gate Avenue
San Francisco, California 94102

Associate Regional Solicitor
U.S. Department of Labor
Federal Building, Room 7725
300 North Los Angeles Street
Los Angeles, California 90012

Region 10—Seattle

Associate Regional Solicitor
U.S. Department of Labor
1911 Smith Tower Building
Seattle, Washington 98104

U.S. DEPARTMENT OF LABOR, BUREAU OF LABOR STATISTICS: REGIONAL OFFICES

Region 1—Boston

Regional Director
Bureau of Labor Statistics
1603-A Federal Office Building
Boston, Massachusetts 02203

Region 2—New York

Regional Director
Bureau of Labor Statistics
1515 Broadway
New York, New York 10036

Region 3—Philadelphia

Regional Director
Bureau of Labor Statistics
Penn Square Building, Room 406
1317 Filbert Street
Philadelphia, Pennsylvania 19107

Region 4—Atlanta

Regional Director
Bureau of Labor Statistics
1371 Peachtree Street, N.E.
Atlanta, Georgia 30309

Region 5—Chicago

Regional Director
Bureau of Labor Statistics
300 South Wacker Drive, 8th Floor
Chicago, Illinois 60606

Region 6—Dallas

Regional Director
Bureau of Labor Statistics
1100 Commerce Street, Room 6B7
Dallas, Texas 75202

Regions 7 and 8—Kansas City and Denver

Regional Director
Bureau of Labor Statistics
Federal Office Building
911 Walnut Street
Kansas City, Missouri 64106

Regions 9 and 10—San Francisco and Seattle

Regional Director
Bureau of Labor Statistics
450 Golden Gate Avenue
Box 36017
San Francisco, California 94102

PUBLICATIONS AND PERIODICALS

The following list of publications and periodicals should not be interpreted as the only literature available on the subjects of safety and health. These are, however, some major publications available for public use.

Subject: Fire Protection and Control

Title	Publisher
Fire Engineering	Reuben H. Donnelley 466 Lexington Avenue New York, New York 10017

Fire Journal
Fire News
Firemen
Fire Technology

National Fire Protection
 Association
60 Batterymarch Street
Boston, Massachusetts 02110

Subject: Hazards

Occupational Hazards

Industrial Publishing
 Corporation
812 Huron Road
Cleveland, Ohio 44115

Subject: Health

AMA Archives of Environmental Health

American Medical Association
535 North Dearborn Street
Chicago, Illinois 60610

American Industrial Hygiene Association Journal

American Industrial Hygiene
 Association
210 Haddon Avenue
Westmont, New Jersey 08108

Chemical Abstracts (Toxicology, Air Pollution, and Industrial Hygiene Section)

American Chemical Society
1155 Sixteenth Street, N.W.
Washington, D.C. 20036

Industrial Hygiene News Report

Flournoy and Associates
1845 West Morse Avenue
Chicago, Illinois 60626

Industrial Hygiene Digest

Industrial Hygiene Foundation
5231 Centre Avenue
Pittsburgh, Pennsylvania 15232

Subject: Safety

Professional Safety

American Society of Safety
 Engineers
850 Busse Highway
Park Ridge, Illinois 60068

Safety Standards	Bureau of Labor Statistics U.S. Department of Labor Washington, D.C. 20210
National Safety News *Industrial Supervisor* *Traffic Safety* *Journal of Safety Research* *Safe Worker* *Safe Driver* *Safety Newsletters*	The National Safety Council 425 North Michigan Avenue Chicago, Illinois 60611

SPECIFIC SOURCES OF SAFETY INFORMATION AND DATA ANALYSIS

American Medical Association, *Hygienic Guide Series*, Southfield, MI.

American Management Association, Inc., *Safety for the Supervisor* (Programmed Instruction for Management Education), New York, 1964.

American Society of Safety Engineers, *A Selected Bibliography of Reference Materials in Safety Engineering and Related Fields*, W. H. Tarrants (Ed.), Park Ridge, IL, 1967.

Alfred, M. Best Company, *Best's Safety and Maintenance Directory*, Morristown, NJ, Published annually.

Bird, F. E., Jr., and G. L. Germain, *Damage Control*, New York: American Management Association, Inc. 1966.

Blake, R. P., *Industrial Safety*, 3rd edition, Englewood Cliffs, N.J.: Prentice-Hall, Inc., 1963.

Brown, K. S., and J. B. Re Velle, *Quantitative Methods for Managerial Decisions*, Reading, MA: Addison-Wesley, 1978.

Bureau of National Affairs, Inc., *ABC's of the Job Safety and Health Act*, Washington, D.C., 1971.

Bureau of National Affairs, Inc., *The Job Safety and Health Act of 1970*, Washington, D.C., 1971.

Commerce Clearing House, Inc., *Occupational Safety and Health Act of 1970: Law and Explanation*, Chicago, 1971.

Davidson, R., *Peril on the Job*, Washington, D.C.: Public Affairs Press, 1970.

DeReamer, R., *Modern Safety and Health Technology*, New York: John Wiley & Sons, Inc., 1980.

E. I. Du Pont de Nemours and Co., Inc., *Safety Training Observation Program (S.T.O.P.)*, Wilmington, DE, 1970.

Eninger, M. U., *Accident Prevention Fundamentals for Supervisors and Managers*, Toronto, Ontario: Industrial Accident Prevention Associations, 1968.

Fawcett, H. H., and W. S. Wood, *Safety and Accident Prevention in Chemical Operations*, New York: John Wiley & Sons, Inc. 1965.

Fletcher, J. A., and H. M. Douglas, *Total Environmental Control*, Ontario: Hunter Rose Company, 1970.

Gilmore, C. L., *Accident Prevention and Loss Control*, New York: American Management Association, Inc., 1970.

Grimaldi, J. V., and R. H. Simonds, *Safety Management*, Homewood, IL: Richard D. Irwin, Inc., 1963.

Haddon, W., Jr., "The Prevention of Accidents," in *Preventive Medicine*, D. W. Clark and B. MacMahon, Boston: Little, Brown and Company, 1967.

Heinrich, H. W., *Industrial Accident Prevention*, 4th edition, New York: McGraw-Hill Book Company, Inc., 1959.

Industrial Hygiene Foundation of America, Inc., *Industrial Hygiene Highlights*, Pittsburgh, 1968.

National Fire Protection Association, *Fire Protection Handbook*, 13th edition, Boston, 1969.

National Fire Protection Association, *Inspection Manual*, Boston, 1970.

National Safety Council, *Accident Facts*, Chicago, published annually.

National Safety Council, *Accident Prevention Manual for Industrial Operations*, 7th edition, Chicago, 1969.

National Safety Council, *Supervisors' Safety Manual*, 3rd edition, Chicago, 1967.

Patty, F. A., *Industrial Hygiene and Toxicology*, New York: Interscience Publishers, Inc., 1958.

Pope, W. C., and E. R. Nicolai, *In Case of Accident, Call the Computer*, Washington, D.C.: U.S. Department of Interior, 1971.

Sax, N. I., *Handbook of Dangerous Material*, New York: Reinhold Publishing Company, 1951.

U.S. Department of Health, Education and Welfare, *The Industrial Environment—Its Evaluation and Control*, U.S. Department of Labor, *All About OSHA*, Superintendent of Documents, Washington, D.C.

U.S. Department of Labor, *Code of Federal Regulations, 29 Labor*, Part 900 to End, Washington, D.C.

U.S. Department of Labor, *Federal Register*, Occupational Safety and Health Administration, Vol. 36, No. 105, Washington, D.C., May 29, 1972.

U.S. Department of Labor, *Recordkeeping Requirements Under the Williams-Steiger Occupational Safety and Health Act of 1970*, Washington, D.C., 1978.

U.S. Department of Labor, *Safety in Industry*, Bureau of Labor Standards Training Programs, Washington, D.C.

Index

Abrasive wheel, 172
Accident causes, 101
  "Accident Facts," 187
  classifications, 74
  frequency rate, 185
  investigation: conduct, 121
    inquiry, 118
    investigator, 118
    program responsibility, 121
    recommendations, 120
    records, 75
    when needed, 118
  investigators, 71
  prevention, signs and tags, 47
  procedures, post-, 69
  proneness, 68
  report: computerized, 75
    content, 117
    preparation of, 115
    terminology, 118
  reporting and records system, 148
  severity rate, 187
Acetylamino-Fluorene, 62
Aerial boom(crane) operation, 173
Agreement, OSHA–American Red Cross, 176
Air receivers (boilers), 152
Aisles, 150
  fire, 153
Alarm switches and controls, 150
Alpha-Naphthyl-Amine, 54
American Conference of Governmental Industrial Hygienists, 183
American Medical Association (AMA), 174
American National Standards Institute (ANSI), 74
American Red Cross, 176
  OSHA agreement, 176
Aminodiphenyl, 60
Analysis of variance, 223
Anhydrous Ammonia, 45
ANSI J6.1 to 6.7, 98
  Z16, 74
  Z24.22, 93
  Z41.1, 95
  Z87.1, 85
  Z88.2, 88
  Z89.1, 89
Anticipated results of safety training program, 19
Arc welding, 168
Area rules, 6
Areas for improvement, 7
  high accident, 154
  inspection problem, 155
Atmospheric pressure, 100
Attitude survey, 195
  content variance, 197
  employee, 70
  response rate, 195
Attitudes, employee, 73
Audio visual materials, 41
Audit inspection record, 75
Audit inspections, 158
Authority, 11
Average, arithmetic, 197
Average cost per injury, 190
Awareness of Safety Knowledge (ASK) exam, 204

Barlow case, 136
Barrier creams, 85
Barrier, physical, 170
Basic functions of safety program, 23
Behavior modification, 71
Belt and pulley guarding, 173
Bench grinder, 171
Benzidine, 59
Beta-Naphtyhlamine, 58
Beta-Propiolactone, 61
Bis-Chloromethyl Ether, 57
Black lantern trophy, 9
Blasting agents, 45
Blind persons, 150
Boilers, 152
Brainstorming, 8
Brazing, welding and cutting, 49
Budget, safety program, 148
Building rules, 7
Building, structural integrity, 149
Bulletin board, 9
    safety, 72
Bureau of Labor Statistics (BLS), 78
Buzz saw, 172

Calculations, cost-benefit, 223
Carbon Dioxide, extinguishing systems, 47
Case, medical, 75
Causes of accidents, 101
Centralized operations, 10
Central tendency, 197
Chain, 173
Checklist, safety inspection, 148
    safety meeting, 5
Checks, communication, 70
    spot inspection, 70
Chi-square analysis, 212
Chocked vehicles, 166
Circuit overloading, 151
Classifying accidents, 74
Class size, safety training, 27
Climate, safety, 66
Clothing and equipment, protective, 7
    disposable, 87
Code of Federal Regulations (CFR), 42, 43, 135
Color code, safety, 150
Committee meetings, 148
Communication, 14
    checks, 70
    lines of, 40
    media, 18
Communications, 88
Compliance officers (CO's), 135, 141
    scope, 136
Compressed gas cylinders, 151, 165
Content variance, attitude survey, 197
Contests, departmental, 5
Contigency tables, 212
Contract services, medical, 153
Controls, alarm, 150
    engineering, 169
Construction site, 169
Consulting physician, 181
Correlation analysis, 204
    coefficient, 205
Cost-benefit calculations, 223
Cost effectiveness, 154
Course outline, review of, 21
Creams, barrier, 85
Crane: crawler locomotive and truck, 48
    overhead and gantry, 48
Crane (aerial boom) operation, 173
Cutting, welding and brazing, 49

Dangerous activities, recognition of, 100
Data expansion, for correlation coefficient, 207
Decentralized operations, 10
Defective pallets, 170
Definitions, 224
Degrees of freedom, 213
Demonstration of job, 38
Departmental contests, 5
Dermatitis, 168
Dichloro-Benzidine, 57
Dimethylaminoazo-Benzene, 63
Director of safety, 1
Disabling injury, 75
Discussion topics in safety training, 29
Dispensary, 177
Disposable clothing, 87
Doctors, 153
Doors, 149
    fire, 153

Ear protection, 171
    models, 93
    personal, 169
Electrical devices: maintenance, 133

Index

protection, 98
  equipment, 152
  grounded, 152
  fuses, 152
  lines overhead, 173
  wiring, 151
Electric light bulbs, fire hazard, 132
Elevators, 150
  inspection and maintenance, 152
Elevator shaft, open, 170
Elimination of unsafe conditions, 113
Emergency evacuation plans, 69
  information signs, 71
  response, fire, 125
  switches, 151
Empirical studies, 12
Employee: analysis of, 37
  attitude, 73
  attitude survey, 70
  awareness, 22
  orientation, 34
  personal problems, 14
  resistance to equipment, 80
  safety committee, 4
Empoyees, 184
  analysis of, 37
  needs to know, 72
  new, 73
  safety committee, 4
Engineering controls, 169
  organization, 183
Environmental hazards protection, 99
Equipment: adequacy, 80, 83
  approved, 149
  break-in period, 81
  and clothing, protective, 7
  company policy, 80
  electrical, 152
  electrical, grounded, 152
  employee resistance to, 80
  enforcement, 81
  fitting and adjustment, 81
  follow-through, 81
  hazardous arrangement, 145
  implementing use of, 81
  inspections, 84, 88
  maintaining use of, 81
  maintenance, 80, 83
  need analysis, 81
  personal protective, 79

  photographic, 147
  planning use of, 81
  position, 145
  program communication, 81
  protective, personal, 149, 151, 153
  for safety training, 27
  sanitation, 80, 83
  selection, 81
  shut-down, 69
  target date setting, 81
  training, 81
  training for use of, 80
Ethyleneimine, 60
Evacuation plan, emergency, 69
Evaluation of training program, 22
Executive safety committee, 148
Exit signs, 150
Expected number (value), 214
Explosives, 45
Extension cords, 151
External data, comparisons, 78
Eye and face protection, 84
Eye hazards, 153
Eye protection, 171
Eyewash fountain, 167

Face protection, 171
  shield, 153
Facilities, medical, 177
  grounds, 150
  shutdown, 69
Facilities appearance, 202
Files: medical alert, 75
  safety, 75
Films, safety, 72
Finger cuts, 85
Fire: electrical, 132
  industrial causes, 122
Fire aisles, 153
  department inspection record, 75
  doors, 153
  emergency, response to, 125
  escape stairway, 153
  extinguishers, 153, 164
    application of, 129
    guide to, 128
    hazards in, 129
    instruction in, 127, 153
    location and placement, 130
    maintenance, 127

new instructions, 126
NFPA classifications, 126
  selection, 130
  types, 126, 127
hazards, 152
  electrical, 132
  electric bulbs, 132
prevention program, components, 123
protection, 6
and control publications, 235
First aid, 47, 175
  cards, 72
  facility, 71
  kits, 71, 177
    OSHAct regulations, 180
    physician approval, 178
    placement of, 178
    stocking of, 179
    unit type, advantages of, 178
  room, 153, 177
  training, 174, 176
Flammable and combustible liquids, 45, 152
Flammable materials, 152
Flammables, 145
Flash points, 134
Floors, 149, 164
Follow-up, 151
Foot protection, 95, 171
Foremen, 149
Forging machines, 48
Formal inspections, 159
Form, plant noise survey, 96
Fountain: eyewash, 167
  water, 149
Frequency distribution: Poisson, 212
  population, 212
  sample, 212
Frequency-severity indicator, 188
Fumes, 168
  hazardous, 151
Fuses, electrical, 152

Gantry cranes, 48
General body protection, 87
General plant rules, 6
General requirements, OSHA, 79
Gloves, 85
  rubber insulating, 98
Goals: defined, 7

quantitative terms, 7
Goggles, 153
Goodness-of-fit test, 212
Grading systems, quantitative, 204
Grinder, bench, 171
Grounded, equipment, electrical, 152
Grounds, facility, 150
Ground, soft, 170
Group, commitment, 8
Group meetings, small, 72
Guarding: machine, 85, 150
  mechanical, 166
  saws, 169
Guards: belt and pulley, 173
  mechanical, 162
  point of operation, 151
  spark-deflecting, 171

Handicapped persons, 149
Hand, pads, 85
Hand protection, 85
Hard hats, 9, 169, 171
Hazard, abatement, 140
Hazard inspection record, 75
Hazard, inspection training, 147
Hazardous arrangement equipment, 145
Hazardous fumes, 151
Hazardous gases, 133
Hazard reporting: categorical approach, 200
  combined approaches, 201
  departmental approach, 201
  overall approach, 200
  per capita/departmental approach, 202
  periodic approach, 200
Hazard reporting rate, 198
Hazard reports, 69
Hazards: control, 183
  degrees of severity, 115
  evaluation, 183
  eye, 153
  fire, 152
  identification by personnel, 146
  protruding, 150
  publications, 236
  recognition, 183
  respiratory, 153
Head protection, occupational, 90
Health Publications, 236
Hearing protection, 93

Index

Helmets: protective, 89
  types of, 90
High accident areas, 154
Housekeeping, 142, 152, 163, 166
  areas of concentration, 143
  checklist, 203
  procedures, 142
Human failures, categories, 101
Human relations movement, 12
Hygiene training, industrial, 181

Illness, OSHA recordable, 7
Illumination: levels, 151
  sources, 151
Imminent danger, 140, 224
Industrial fires, causes of, 122
Industial hygiene: stress categories, 182
  training, 181
Industrial Hygienists, American Conference of Governmental, 183
Industrial sink, 168
Industrial trucks, powered, 48
Influencing people, 12
Injury: disabling, 75
  to employees, consequences of, 17
  to employers, 17
  experience, measurement, 77
  lost-time, 75
  minor, 75
  occupational, 74
  off-the-job, 17
  OSHA recordable, 77
  report, model, 115, 116
Inspections: audit, 158
  checklist, 148
  equipment, 88
  formal, 149
  guide, OSHA standards, 159
  no-notice, 71
  objectives, 156
  organizing, 157
  planning, 156
  plant, simulated, 160
  principles, 156
  prior to, 157
  spot checks, 70
  safety, 149
    periodic, 8
  series of, 71
  types of, 157

Instruction, 151
  fire extinguishers, 153
Instruction sheets, 41
Instructor checklist, 28
Instructor's job in safety training, 27
Insurance company inspection record, 75
Internal data analyses, 77
Internal publications, 71
Ionizing radiation, 44

Job demonstration, 38
Job-home gap, 17
Job safety analysis, 35
Job site survey, 223

Ladder, portable, 162
Laundry machinery, 49
Leadership, 148
Leggings, 87
Lesson plan: safety training, 29
  sample, 30
Lifelines, 87
Lifevests, 87
Light, ultraviolet, 168
Line positions, 10
Link, chain, 173
Liquids, flammable and combustible, 45
Location of safety training, 26
Logging, pulpwood, 49
Lost-time injury, 75

Machine demonstration, 39
Machine, guarding, 85
Machinery: guarding, 150
  laundry, 49
Machines: forging, 48
  unsafe, 40
Maintenance program, 149
  chain and link, 173
Management involvement, 1
Management overview, 72
Management statement, 4
Manager, plant, 148
Materials, stored, 150
Matrix, 216
Matrix organiztion, 12
Mean: arithmetic, 197
  population, 213
  sample, 213
Measurement: injury experience, 77

noise, 169
Measures of: association, 212
  dispersion, 197
  safety, 185
Mechanical power presses, 22, 48
Media for promoting safety, 18
Medical alert file, 75
  case, 75
  facilities, 177
  personnel, 183
  program, 153
  records, 177
    standardization, 177
  services, 47
    contract, 153
  training, 174
    OSHAct standards, 180
Meeting minutes, 148
Meeting place, safety class, 27
Meetings: committee, 148
  safety committee, guests, 72
  small group, 72
Methyl Chloro-Methyl Ether, 56
Methylene Bis (2-Chloro-Aniline), 55
Minutes of meetings, 148
Model injury report, 115, 116
Model safety program, 225
Modes of entry, body, 182
Motivation, 15
Motivational training, 80
Motivation for safety, 31

National Electric Code (NEC), 147, 151, 152
National Safety Council, 187
Need to know, employees, 72
Newsletter, 2
Nitrobiphenyl, 53
Nitrosodi-Methyl-amine, 64
Noise: area, 168
  exposure, 95
  intensity, 169
  measurement, 169
  suppression/containment, 71
  survey form, 96
No-notice inspection, 71
  record, 75
"No Smoking" signs, 152
Null hypothesis, 212

Nurse, registered, 153

Observed number, 215
Occupational health, 167
Occupational injuries, 74
Off-the-job injuries, 17
Old habits, 73
Open pit, 162
Organizational structure, 10
Organization: matrix, 12
  responsibilities of, 184
  traditional, 12
Orientation, employees, 34
  program, 228
  topics for, 34
OSHA-American Red Cross, agreement, 176
OSHA: appeal procedures, 140
  appeals, 140
  citations, 140
  compliance program, 147
  general requirements, 79
  penalties, 140
  posting of citations, 140
  standards, 147
    abstract, 160, 161
    as inspection guide, 159
    kinds of, 137
    training requirements, 73
  violations, types of, 141
OSHA inspection, prior to, 138
OSHA medical training standards, 180
OSHA priority hazards, 104
OSHA recordable illness, 77
OSHA recordable injury, 77
OSHA reports, submission requirements, 117
OSHA standards violations: abrasive wheel adjustment, 107
  blocked or inadequate exits, 110
  compressed gases, 107
  electrical, 104
  eye and face protection, 111
  fire extinguishers, 105
  fixed ladders, 112
  flammable and combustible liquids, 109
  hand and portable power tools, 110
  mechanical guarding, 106
  mechanical guards, 104

Index

medical services and first aid, 111
noise exposure, 111
personal protective equipment, 112
portable ladders, 110
powered industrial trucks, 108
sanitation, 109
saw guarding, 106
spray finishing, 109
unmarked exits, 108
walking and working surfaces, 106
OSHA training standards, 42
OSHAct regulations, first aid kit, 180
OSHAct: requirements, 135
  training responsibilities, 139
OSHA 101, 75
OSHA 200, 75
OSHA 2201, 70, 147
OSHA 2213, 224, 156
OSHA 2215, 154
OSHA 2254, 42, 140
Overhead cranes, 48
Overhead lines, electrical, 173
Overloading of circuits, 151

Pallets, defective, 170
Parking lots, 150
Periodic review of rules, 7
Periodic safety inspections, 8
Personality factors, 13
Personal problems, employees, 14
Personal protection, ear, 169
Personal protective equipment, 79, 149, 153
Personnel: assignment, 175
  evaluation ratings, 203
  hazard identification by, 146
  productivity indicators, 203
  selection, 175
  training, 175
Photographic equipment, 147
Physical barrier, 170
Physicals, preemployment, 86
Physician, consulting, 181
Physician approval, first aid kit, 178
Pit, open, 162
Planning inspections, 156
Plans, emergency evacuation, 69
Plant, general rules, 6
Plant manager, 148

Plant noise survey form, 96
Plant organization factors, 113
Pledge card, safety, 9
Point of operation guards, 151
Poisson frequency distribution, 212
(Poly) Vinyl Chloride, 64
Population frequency distribution, 212
Post accident procedures, 69
Posters, 18, 41,
  safety, 72
Powered industrial trucks, 22, 48
Power presses, mechanical, 48
Power tools, 151
Preemployment physical, 86
Prepare the learner, 38
Presentation of safety training, 27
Press, punch, 168
Pressure atmospheric, 100
Pressure vessels, 152
Priority hazards, OSHA, 104
Probabilistic analysis, 219
Procedures: post accident, 69
  special, safety-related, 69
Program, behavior modification, 71
Program requirements, respiratory
  protection, 88
Project, noise suppression containment, 71
Promotion, safety program, 14
Promotion media, 18
Protection: ear, eye or foot, 171
  ear, models, 93
  electrical device, 98
  eye and face, 84, 171
  foot, 95
  general body, 87
  gloves, 85
  hand, 85
  hazards, environmental, 99
  head, 90
  hearing, 93
  personal, ear, 169
  respiratory, 46, 87
Protective clothing and equipment, 7
Protective equipment, personal, 79, 149, 151, 153
Protective helmets, 89
  types, 90
Protective shoe, classes, 97
Protruding hazards, 150

Publications, Fire Protection and Control, 235
  hazards, 236
  health, 236
  internal, 71
Publications: and periodicals, 235
  safety, 236
Pulpwood logging, 49
Punch press, 168
Purchasing organization, 183

Quantitative grading systems, 204

Radial saw, 169
Ramps, 149
Rate: accident frequency, 185
  accident severity, 187
  hazard reporting, 198
Recognition, dangerous activities, 100
Records: accident investigation, 75
  anecdotal, 40
  audit inspection, 75
  fire department inspection, 75
  hazard inspection, 75
  insurance company inspections, 75
  maintenance, 75
  master file, 75
  medical, 77
    standardization, 177
  no-notice inspection, 75
  OSHA requirements, 77
  safety, 5
  training, 16
  training program, 75
Reference sources, 237
Registered nurse, 153
Regression analysis, 208
Regression line: slope, 208
  Y-intercept, 208
Rehabilitation program, employees:
    alcoholics, 175
  drug problem, 175
Report of accident: computerized, 75
  content, 117
  terminology, 118
Reporting form, 69
Report of injury, model, 115, 116
Report preparation, accident, 115
Report of safety self-inspection, 154
Reports: hazard, 69

requirements for submission, OSHA, 117
  safety inspection, 40
Report, uniform terminology, 119, 120
Respirators, 88
Respiratory hazards, 153
Respiratory protection, 46, 87
  program requirements, 88
Response rate, attitude survey, 195
Responsibility, 12
Review course outline, 21
Rubber insulating gloves, 98
Rubber insulating sleeves, 98
Rules: area, 6
  building, 7
  general plant, 6
  periodic review of, 7
  safety training, 37
  specific, 7

Safe-T-Score, 192
Safety box, 10
Safety, bulletin boards, 72
Safety climate, 66, 148
Safety: clubs, 83
  color code, 150
  films, 72
  glasses, 84
  goggles, 85
  and health training, 4 R's, 24, 25
  incentives, 230
  inspection, 40, 149
    checklist, 148
    periodic, 8
    reports of, 40
  leaders, departmental, 4
  measures, 185
  media for promoting, 18
  meeting: checklist, 5
    departmental, 4
  motivation, 31
  observation tours, 147
  performance indicators, 203
  pledge, 40
    card, 9
Safety committee, 4
  employee, 4
  employer, 4
  executive, 148
  meeting, guests at, 72

Index

"Safety Day," 72
Safety department, 184
Safety design, 68
Safety files, requirements, 76
  policy, 1, 148
  posters, 72
  practices and procedures, 40
  program, 148
    AMA objectives, 174
    basic functions, 23
    budget, 148
    incentives, 230
    model, 225
    promotion, 14
    teaching aids, 40
  publications, 236
  recommendation, 40
  records, 5
  regulation, individual development, 40
  sampling, 223
  self-inspection report, 154
  tests, 40
  training: amount of, 216
    class size, 27
    discussion topics, 29
    equipment, 27
    evaluation of, 185
    for foreman, 149
    instructor checklist, 28
    instructor's job, 27
    lesson plans, 29
    location of, 26
    meeting place, 27
    presentation of, 27
    rules for, 37
    sequence of, 39
    step of, 39
    tests, 29
    timing of, 26
    visual aids, 29
    who conducts, 26
  training program, 15
    anticipated results, 19
    characteristics, 39
Safety working conditions, 40
Sample frequency distribution, 212
Sample lesson plan, 30
Saw: buzz, 172
  radial, 169
Saws, guarding of, 169
  unguarded, 172
Scaffolding, 172
Scientific Management, 68
SCRAPE, 223
Secretary of Labor, 176
Self-appraisal, 33
Self-inspection guide, 179
Self-inspection report, safety, 154
Series of successes, 67
Service organizations and associations, 232
Set example, 40
Shaft, elevator, open, 170
Shutdown, facilities and equipment, 69
Signals: for training, 73
  warning, 150
Signs: emergency information, 71
  exit, 150
  "No Smoking", 152
  and tags, for accident prevention, 47
  warning, 170
Simulated plant inspection, 160
Sink, industrial, 168
Sleeves, 87
  rubber insulating, 98
Slings, 167, 171
Slope, of regression line, 208
Small group meetings, 72
Smokers, 124, 145
Special Computerized Accident Report (SCAR), 75
Special procedures, 69
Specific rules, 7
"Spotlight on Safety," 71
Spray painting, 152
Sprinkler heads, 153
Stacking, 170
Staff positions, 10
Staff responsibility, 148
Stairs, 149, 163
Stairway, fire escape, 153
Standard deviation, 197
Standard Industrial Code (SIC), 78
Statistical analysis, 193
Statistical association, 219
  lack of, 212
Statistical control charts, 203
Statistical techniques, 204
Statistics, 74
  Bureau of Labor, 78

Steps, 149
Storage areas, 150
Stored materials, 150
Stress: categories, industrial hygiene, 182
  symptoms of, 15
Student's t, 193
Suggestion box, 41
Suggestion system, 8
Supervision, 12
  equipment operation, 39
Supervisor, 184
  actions, 69
  cooperation with, 67
  first-line, 66
  set example by, 66
  unsafe practices by, 103
Supervisory functions 12,
  ratings, 216
Supreme Court rulings, 136
Survey, employee attitude, 70
Switches: alarm, 150
  emergency, 151

Teaching aids, safety program, 40
Temperature, extremes, 99
Test: data independence, 212
  goodness-of-fit, 212
  and quizzes, 41
  safety training, 29
Threshold Limit Values (TLV's), 182
Timing of safety training, 26
Toilet areas, 149
Toxic substances, modes of entry, body, 182
Traditional organization, 12
Training: amount of, 216
  areas, 21
  first aid, 174, 176
  hazard inspection, 147
  industrial hygiene, 181
  medical, 174
  motivational, 80
  OSHAct standards, medical, 180
  programs, 21
    evaluation of, 22
    records, 75
  records of, 16

  requirement, OSHA standards, 42, 73
  safety, evaluation of effectiveness, 185
  signals, 73
  specialized, 16
Trucks, powered industrial, 48

Ultraviolet light, 168
Unauthorized flammables, 145
Unguarded saw, 172
Unit type kits, first aid, advantages of, 177
Unsafe acts, 100
  categories, 101
  response to, 101
Unsafe building conditions, 144
Unsafe conditions, 144
  checklist, 114
  descriptions, 115
  elimination of, 113
Unsafe machinery and equipment, 145
Unsafe machines, 40
Unsafe practices, supervisors, 103
Unsupervised work, 40
U. S. Department of Labor, Bureau of Labor Statistics, 235
  Office of the Soliciter, 233

Vehicles, 150
Ventilation, 43, 151
Vinyl Chloride (Poly), 64
Violations, housekeeping, 142
Visual aids for safety training, 29
Vortex suit, 100

Warning signals, 150
Warning signs, 170
Waste containers, 152
Waste paper and rags, 152
Water fountain, 149
Welding: arc, 168
  cutting and brazing, 49
Wheel, abrasive, 172
Who conducts safety training, 26
Workmen's compensation premium ratings, 203
Worst-first, 154

Y-intercept, of regression line, 208